INDUSTRY 4.0

THE INDUSTRIAL INTERNET OF THINGS

Alasdair Gilchrist

Apress®

Industry 4.0: The Industrial Internet of Things

Alasdair Gilchrist
Bangken, Nonthaburi
Thailand

ISBN-13 (pbk): 978-1-4842-2046-7   ISBN-13 (electronic): 978-1-4842-2047-4

DOI 10.1007/978-1-4842-2047-4

Library of Congress Control Number: 2016945031

Copyright © 2016 by Alasdair Gilchrist

This work is subject to copyright. All rights are reserved by the Publisher, whether the whole or part of the material is concerned, specifically the rights of translation, reprinting, reuse of illustrations, recitation, broadcasting, reproduction on microfilms or in any other physical way, and transmission or information storage and retrieval, electronic adaptation, computer software, or by similar or dissimilar methodology now known or hereafter developed. Exempted from this legal reservation are brief excerpts in connection with reviews or scholarly analysis or material supplied specifically for the purpose of being entered and executed on a computer system, for exclusive use by the purchaser of the work. Duplication of this publication or parts thereof is permitted only under the provisions of the Copyright Law of the Publisher's location, in its current version, and permission for use must always be obtained from Springer. Permissions for use may be obtained through RightsLink at the Copyright Clearance Center. Violations are liable to prosecution under the respective Copyright Law.

Trademarked names, logos, and images may appear in this book. Rather than use a trademark symbol with every occurrence of a trademarked name, logo, or image we use the names, logos, and images only in an editorial fashion and to the benefit of the trademark owner, with no intention of infringement of the trademark.

The use in this publication of trade names, trademarks, service marks, and similar terms, even if they are not identified as such, is not to be taken as an expression of opinion as to whether or not they are subject to proprietary rights.

While the advice and information in this book are believed to be true and accurate at the date of publication, neither the authors nor the editors nor the publisher can accept any legal responsibility for any errors or omissions that may be made. The publisher makes no warranty, express or implied, with respect to the material contained herein.

> Managing Director: Welmoed Spahr
> Lead Editor: Jeffrey Pepper
> Technical Reviewer: Ahmed Bakir
> Editorial Board: Steve Anglin, Pramila Balan, Louise Corrigan, James T. DeWolf, Jonathan Gennick, Robert Hutchinson, Celestin Suresh John, James Markham, Susan McDermott, Matthew Moodie, Ben Renow-Clarke, Gwenan Spearing
> Coordinating Editor: Mark Powers
> Copy Editor: Kezia Endsley
> Compositor: SPi Global
> Indexer: SPi Global
> Artist: SPi Global

Distributed to the book trade worldwide by Springer Science+Business Media New York, 233 Spring Street, 6th Floor, New York, NY 10013. Phone 1-800-SPRINGER, fax (201) 348-4505, e-mail orders-ny@springer-sbm.com, or visit www.springeronline.com. Apress Media, LLC is a California LLC and the sole member (owner) is Springer Science + Business Media Finance Inc (SSBM Finance Inc). SSBM Finance Inc is a Delaware corporation.

For information on translations, please e-mail rights@apress.com, or visit www.apress.com.

Apress and friends of ED books may be purchased in bulk for academic, corporate, or promotional use. eBook versions and licenses are also available for most titles. For more information, reference our Special Bulk Sales–eBook Licensing web page at www.apress.com/bulk-sales.

Any source code or other supplementary materials referenced by the author in this text is available to readers at www.apress.com/9781484220467. For detailed information about how to locate your book's source code, go to www.apress.com/source-code/. Readers can also access source code at SpringerLink in the Supplementary Material section for each chapter.

*To my beautiful wife and daughter,
Rattiya and Arrisara, with all my love*

Contents

About the Author vii
About the Technical Reviewer ix
Acknowledgments xi
Introduction xiii

Chapter 1: Introduction to the Industrial Internet 1
Chapter 2: Industrial Internet Use-Cases 13
Chapter 3: The Technical and Business Innovators of the
           Industrial Internet 33
Chapter 4: IIoT Reference Architecture 65
Chapter 5: Designing Industrial Internet Systems 87
Chapter 6: Examining the Access Network Technology
           and Protocols 119
Chapter 7: Examining the Middleware Transport Protocols 125
Chapter 8: Middleware Software Patterns 131
Chapter 9: Software Design Concepts 143
Chapter 10: Middleware Industrial Internet of Things Platforms .. 153
Chapter 11: IIoT WAN Technologies and Protocols 161
Chapter 12: Securing the Industrial Internet 179
Chapter 13: Introducing Industry 4.0 195
Chapter 14: Smart Factories 217
Chapter 15: Getting From Here to There: A Roadmap 231

Index 245

About the Author

Alasdair Gilchrist has spent his career (25 years) as a professional technician, manager, and director in the fields of IT, data communications, and mobile telecoms. He therefore has knowledge in a wide range of technologies, and he can relate to readers coming from a technical perspective as well as being conversant on best business practices, strategies, governance, and compliance. He likes to write articles and books in the business or technology fields where he feels his expertise is of value. Alasdair is a freelance consultant and technical author based in Thailand.

About the Technical Reviewer

Ahmed Bakir is the founder and lead developer at devAtelier LLC (www.devatelier.com), a San Diego-based mobile development firm. After spending several years writing software for embedded systems, he started developing apps out of coffee shops for fun. Once the word got out, he began taking on clients and quit his day job to work on apps full time. Since then, he has been involved in the development of over 20 mobile projects, and has seen several enter the top 25 of the App Store, including one that reached number one in its category (Video Scheduler). His clients have ranged from scrappy startups to large corporations, such as Citrix. In his downtime, Ahmed can be found on the road, exploring new places, speaking about mobile development, and still working out of coffee shops.

Acknowledgments

Initially, I must thank Jeffery Pepper and Steve Weiss of Apress for their patience and dedication, as the book would not have come to fruition without their perseverance and belief. Additionally, I have to thank Mark Powers for his project management skills and Matt and Ahmed for their technical editing skills. Matt's and Ahmed's editing has transformed the book and for that, I thank you. I would also like to acknowledge my agent Carole Jelen for introducing me to Apress; I cannot thank you enough. Finally, I acknowledge the tolerance of my wife and daughter who complained about the time I hogged the computer and Internet much while writing this book.

Introduction

Industry 4.0 and the Industrial Internet of Things (IIoT) has become one of the most talked about industrial business concepts in recent years. However, Industry 4.0 and the IIoT are often presented at a high level by consultants who are presenting from a business perspective to executive clients, which means the underlying technical complexity is irrelevant. Consultants focus on business models and operational efficiency, which is very attractive, where financial gains and new business models are readily understandable to their clients. Unfortunately, these presentations often impress and invigorate executives, who see the business benefits but fail to reveal to the client the technical abstraction of the lower-layer complexity that underpin the Industrial Internet.

In this book, we strive to address this failure and although we start with a high-level view of the potential gains of IIoT business incentives and models, and describe successful use-cases, we move forward to understand the technical issues required to build an IIoT network. The purpose is to provide business and technology participants with the information required in deploying and delivering an IIoT network.

Therefore, the structure of the book is that the initial chapters deal with new and innovative business models that arise from the IIoT as these are hugely attractive to business executives. Subsequent chapters address the underpinning technology that makes IIoT possible. As a result, we address the way we can build real-world IIoT networks using a variety of technologies and protocols. However, technology and protocol convergence isn't everything; sometimes we need a mediation service or platform to glue everything together. So for that reason we discuss in the middle chapters protocols, software patterns, and middleware IIoT platforms and how they provide the glue or the looking glass that enables us to connect or visualize our IIoT network.

Finally, we move forward from generic IIoT concepts and principle to Industry 4.0, which relates to industry, and there we see a focus on manufacturing. Industry 4.0 relates to industry in the context of manufacturing, so these chapters consider how we can transform industry and reindustrialize our nations.

CHAPTER 1

Introduction to the Industrial Internet

GE (General Electric) coined the name "Industrial Internet" as their term for the Industrial Internet of Things, and others such as Cisco termed it the Internet of Everything and others called it Internet 4.0 or other variants. However, it is important to differentiate the vertical IoT strategies (see Figure 1-1), such as the consumer, commercial, and industrial forms of the Internet from the broader horizontal concept of the Internet of Things (IoT), as they have very different target audiences, technical requirements, and strategies. For example, the consumer market has the highest market visibility with smart homes, personal connectivity via fitness monitors, entertainment integrated devices as well as personal in-car monitors. Similarly, the commercial market has high marketability as they have services that encompass financial and investment products such as banking, insurance, financial services, and ecommerce, which focus on consumer history, performance, and value. Enterprise IoT on the other hand is a vertical that includes small-, medium-, and large-scale businesses. However this book focuses on the largest vertical

© Alasdair Gilchrist 2016
A. Gilchrist, *Industry 4.0*, DOI 10.1007/978-1-4842-2047-4_1

of them all, the Industrial Internet of Things, which encompasses a vast amount of disciplines such as energy production, manufacturing, agriculture, health care, retail, transportation, logistics, aviation, space travel and many more.

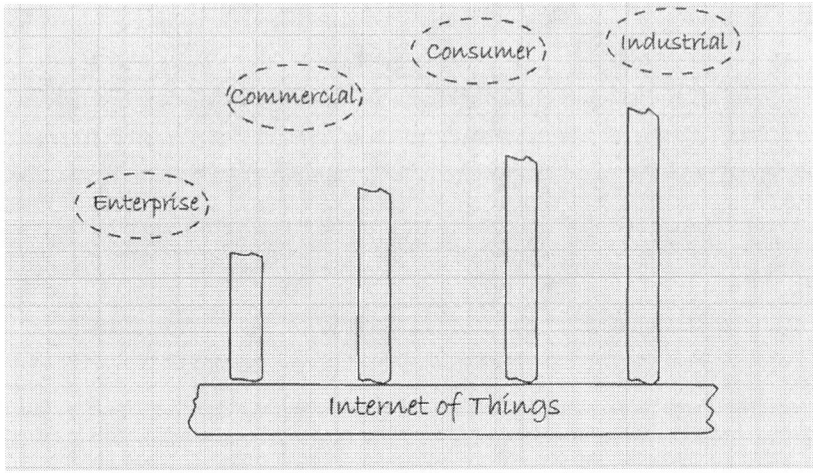

Figure 1-1. Horizontal and vertical aspects of the Internet of Things

In this book to avoid confusion we will follow GE's lead and use the name Industrial Internet of Things (IIoT) as a generic term except where we are dealing with conceptually and strategically different paradigms, in which case it will be explicitly referred to by its name, such as Industry 4.0.

Many industrial leaders forecast that the Industrial Internet will deliver unprecedented levels of growth and productivity over the next decade. Business leaders, governments, academics, and technology vendors are feverishly working together in order to try to harness and realize this huge potential.

From a financial perspective, one market research report forecasts growth of $151.01 billion U.S. by 2020, at a CAGR of 8.03% between 2015 and 2020. However, in practical terms, businesses also see that industrial growth can be realized through utilizing the potential of the Internet. An example of this is that manufacturers and governments are now seeing the opportunity to reindustrialize and bring back onshore, industry, and manufacturing, which had previously been sent abroad. By encouraging reindustrialization, governments hope to increase value-add from manufacturing to boost their GDPs.

The potential development of the Industrial Internet is not without precedence, as over the last 15 years the business-to-consumer (B2C) sector via the Internet trading in retail, media, and financial services has witnessed stellar growth. The success of B2C is evident by the dominance of web-scale giants born on the Internet, such as Amazon, Netflix, eBay, and PayPal. The hope is that the next decade will bring the same growth and success to industry, which in this context covers manufacturing, agriculture, energy, aviation, transportation, and logistics. The importance of this is undeniable as industry produces two-thirds of the global GDP, so the stakes are high.

The Industrial Internet, however, is still in its infancy. Despite the Internet being available for the last 15 years, industrial leaders have been hesitant to commit. Their hesitance is a result of them being unsure as to how it would affect existing industries, value chains, business models, workforces, and ultimately productivity and products. Furthermore, in a survey of industry business leaders, 87% claimed in January 2015 that they still did not have a clear understanding of the business models or the technologies.

This is of course to be expected as the Industrial Internet is so often described at such a high level it often decouples the complexities of the technologies that underpin it to an irrelevance. For example, in industrial businesses, they have had sensors and devices producing data to control operations for decades. Similarly, they have had machine-to-machine (M2M) communications and collaboration for a decade at least so the core technologies of the Industrial Internet of Things are nothing new. For example, industry has also not been slow in collecting, analyzing, and hoarding vast quantities of data for historical, predictive, and prescriptive information. Therefore the question industrial business leaders often ask is, "why would connecting my M2M architecture to the Internet provide me with greater value?"

What Is the Industrial Internet?

To explain why businesses should adopt the Industrial Internet, we need to first consider what the IIoT actual is all about. The Industrial Internet provides a way to get better visibility and insight into the company's operations and assets through integration of machine sensors, middleware, software, and backend cloud compute and storage systems. Therefore, it provides a method of transforming business operational processes by using as feedback the results gained from interrogating large data sets through advanced analytics. The business gains are achieved through operational efficiency gains and accelerated productivity, which results in reduced unplanned downtime and optimized efficiency, and thereby profits.

Although the technologies and techniques used in existing machine-to-machine (M2M) technologies in today's industrial environments may look similar to the IIoT, the scale of operation is vastly different. For example, with Big Data in IIoT systems, huge data streams can be analyzed online using cloud-hosted advanced analytics at wire speed. Additionally, vast quantities of data can be stored in distributed cloud storage systems for future analytics performed in batch formats. These massive batch job analytics can glean information and statistics, from data that would never previously been possible because of the relatively tiny sampling pools or simply due to more powerful or refined algorithms. Process engineers can then use the results of the analytics to optimize operations and provide the information that the executives can transform to knowledge, in order to boost productivity and efficiency and reduce operational costs.

The Power of 1%

However, an interesting point with regard to the Industrial Internet is what is termed the power of 1%. What this relates to is that operational cost/inefficiency savings in most industries only requires Industrial Internet savings of 1% to make significant gains. For example, in aviation, the fuel savings of 1% per annum relates to saving $30 billion. Similarly, 1% fuel savings for the gas-fired generators in a power station returns operational savings of $66 billion. Furthermore, in the Oil and Gas industry, the reduction of 1% in capital spending on equipment per annum would return around $90 billion. The same holds true in the agriculture, transportation, and health care industries. Therefore, we can see that in most industries, a modest improvement of 1% would contribute significantly to the return on investment of the capital and operational expenses incurred by deploying the Industrial Internet. However, which technologies and capital expenses are required when initiating an IIoT strategy?

Key IIoT Technologies

The Industrial Internet is a coming together of several key technologies in order to produce a system greater than the sum of its parts. The latest advances in sensor technologies, for example, produce not just more data generated by a component but a different type of data, instead of just being precise (i.e., this temperature is 37.354 degrees). sensors can have self-awareness and can even predict their remaining useful life. Therefore, the sensor can produce data that is not just precise, but predictive. Similarly, machine sensors through their controllers can be self-aware, self-predict and self-compare. For example, they can compare their present configuration and environment settings with preconfigured optimal data and thresholds. This provides for self-diagnostics.

Sensor technology has reduced dramatically in recent years in cost and size. This made the instrumentation of machines, processes, and even people financial and technically feasible.

Big Data and advanced analytics as we have seen are another key driver and enabler for the IIoT as they provide for historical, predictive, and prescriptive analysis, which can provide insight into what is actually happening inside a machine or a process. Combined with these new breed of self-aware and self-predicting components analytics can provide accurate predictive maintenance schedules for machinery and assets, keeping them in productive service longer and reducing the inefficiencies and costs of unnecessary maintenance. This has been accelerated by the advent of cloud computing over the last decade whereby service providers like AWS provide the vast compute, storage, and networking capabilities required for effective Big Data at low cost and on a pay-what-you-use basis. However, some risk-adverse companies may prefer to maintain a private cloud, either on their own data centers or in a private cloud.

Why Industrial Internet and Why Now?

To comprehend why the Industrial Internet is happening today, when its technologies have been around for a while, we need to look at legacy system capabilities and inefficiencies.

One assumption is that the complexity of industrial systems has outpaced the human operator's ability to recognize and address the efficiencies, thus making it harder to achieve improvements through traditional means. This can result in machines operating well below their capabilities and these factors alone are creating the operational incentives to apply new solutions.

Furthermore, IT systems can now support widespread instrumentation, monitoring, and analytics due to a fall in the costs of compute, bandwidth, storage, and sensors. This means it's possible to monitor industrial machines on a larger scale. Cloud computing addresses the issues with remote data storage; for example, the cost and capacity required to store big data sets. In addition, cloud providers are deploying and making available analytic tools that can process massive amounts of information. These technologies are maturing and becoming more widely available, and this appears to be a key point. The technologies have been around for a while and have been adopted by IT—cloud adaptation and SaaS are prime examples of this. However, it is only recently that industrial business leaders have witnessed the stability and maturity of solutions, tools, and applications within these IT sectors reach a level of confidence and lessen concerns.

Similarly, the maturity and subsequent growth in networks and evolving low-power radio wireless wide area networks (WWAN) solutions have enabled remote monitoring and control of assets, which previously were simply not economical or reliable enough. Now these wireless radio networks have reached a price point and a level of maturity and reliability that works in an industrial environment. Together these changes are creating exciting new opportunities when applied to industrial businesses, machines, fleets, and networks.

The decline in the cost of compute, storage, and networks is a result of the cloud-computing model, which allows companies to gather and analyze much larger amounts of data than ever before. This alone makes the Industrial Internet an attractive alternative to the exclusive M2M paradigm.

However, the Industrial Internet has its own issues, which may well act as severe countermeasures to adoption. These are termed *catalysts* and *precursors* to a successful Industrial Internet deployment.

Catalysts and Precursors of the IIoT

Unfortunately, there are several things an IIoT candidate business simply must have in place before embarking on a serious deployment, discussed in the following sections.

Adequately Skilled and Trained Staff

This is imperative if you expect to benefit from serious analytics work as you will certainly need skilled data scientists, process engineers, and electro-mechanical engineers. Securing talent with the correct skills is proving to be a daunting task as colleges and universities seem to be behind the curve and are still pushing school leavers into careers as programmers rather than data scientists. This doesn't seem to be changing anytime soon. This is despite the huge demand for data scientists and electro-mechanical engineers predicted over the next decade. The harsh financial reality is that the better the data analytical skills, the more likely the company can produce the algorithms required to distil information from their vast data lakes. However, this is not just any information but information that returns true value, aligned to the business strategy and goals. That requires data scientists with expert business knowledge regarding the company strategy and short-medium-long term goals. This is why there is a new C-suite position called the Chief Data Officer.

Commitment to Innovation

A company adopting IIOT has to make a commitment to innovation, as well as taking a long-term perspective to the IIoT project's return on investments. Funding will be required for the capital outlay for sensors, devices, machines, and systems. Funding and patience will be required as performing the data capture and configuring the analytics' parameters and algorithms might not result in immediate results; success may take some time to realize. After all, statistical analysis does not always return the results that you may be looking for. It is important to ask the correct questions. Data scientists can look at the company strategy and align the analysis—the questions of data pools—to return results that align with the company objectives.

A Strong Security Team Skilled in Mitigating Vulnerabilities in Industrial and IT Networks

This is vital, as the IIoT is a confluence of many technologies and that can create security gaps unless there is a deep understanding of the interfaces and protocols deployed. Risk assessments should reveal the most important assets and the highest risk assets and strategic plans developed to mitigate the risk. For example, in a traditional industrial production factory the machines that produce the products such as lathes that operate on programmable templates contain all the intellectual and design knowledge to construct the product. Additionally, security teams should enforce policy and procedures across the entire supply chain.

Innovation and the IIoT

Proponents of the Industrial Internet refer to it as being the third wave of innovation. This is in regard to the first wave of innovation being the industrial revolution and the second wave the Internet revolution. The common belief is that the third wave of innovation, the Industrial Internet revolution, is well under way. However, if it is, we are still in its infancy as the full potential of the digital Internet technology has yet to be realized broadly across the industrial technology sectors. We are beginning to see intelligent devices and intelligent systems interfacing with industrial machines, processes, and the cloud, but not on an industry-wide scale. Certainly, there is not the level of standardization of protocols, interfaces, and application that will undoubtedly be required to create an IIoT value chain. As an example of this, there is currently a plethora of communication and radio protocols and technologies, and this has come about as requirements are so diverse.

In short, no one protocol or technology can meet all use-case requirements. The existence of diverse protocols and technologies makes system integration within an organization complex but with external business partners, the level of complexity can make integrating systems impractical. Remember that even the largest companies in the world do not have the resources to run their own value chains. Therefore, until interfaces, protocols, and applications are brought under some level of standardization, interconnecting with partners will be a potentially costly, inefficient, and possibly an insecure option.

Intelligent Devices

We are witnessing innovation with the development of intelligent devices, which can be new products or refitted and upgraded machinery. The innovation is currently directed toward enabling intelligent devices. This is anything that we connect with instrumentation, for example, sensors, actuators, engines, machines, components, even the human body, among a myriad of other possible items. This is because it is easy and cost effective to add instrumentation to just about any object about which we wish to gather information.

The whole point of intelligent devices in the Industrial Internet context is to harvest raw data and then manage the data flow, from device to the data store, to the analytic systems, to the data scientists, to the process, and then back to the device. This is the data flow cycle, where data flows from intelligent devices, through the gathering and analytical apparatus before perhaps returning as control feedback into the device. It is within this cycle where data scientists can extract prime value from the information.

Key Opportunities and Benefits

Not unexpectedly, when asked which key benefits most IIoT adopters want from the Industrial Internet, they say increased profits, increased revenue flows, and lower operational expenditures, in that order. Fortunately, using Big Data to reap the benefits of analytics to improve operational processes appears to be akin to picking the low hanging fruit; it's easily obtainable. Typically, most industrial companies head straight for the predictive maintenance tactic as this ploy returns the quickest results and return on investment.

Some examples of this are the success experienced by Thames Water, the largest fresh-drinking water and water-waste recycler in the UK. It uses the IIoT for remote asset management and predictive maintenance. By using a strategy of sensors, remote communication, and Big Data analytics, Thames Water can anticipate equipment failures and respond quicker to any critical situation that may arise due to inclement weather.

However, other industries have other tactical priorities when deploying IIoT, one being health and safety. Here we have seen some innovative projects from using drones and autonomous vehicles to inspect Oil and Gas lines in inhospitable areas to using autonomous mining equipment. Indeed Schlumberger is currently using an autonomous underwater vehicle to inspect sub-sea conditions. The unmanned vehicle travels around the ocean floor and monitors conditions for anything up to a year powered only by wave motion, which makes deployment in remote ocean locations possible, as they are both autonomous and self-sufficient requiring no local team support.

Submersible ROV (remote operational vehicles) previously had to be lowered and supported via a umbilical cord from a mother ship on the surface that supplied power and control signals. However, with autonomous ROVs, support vessels no longer have to stay in the vicinity as the ROVs are self powered. Furthermore there is no umbilica3l cord that is susceptible to snagging on obstacles on the seabed.

It is not just traditional industry that can benefit from the Industrial Internet of Things. Health care is another area that has its own unique perspective and targets. In health care, the desire is to improve customer care and quality service. The best metric for a health care company to be judged is how long their patients survive in their tender care, so this is their focus—improving patient care. This is necessary, as hospital errors are still a leading cause of preventable death. Hospitals can utilize miniaturized sensors, such as Google and Dexcoms' initiative to develop disposable, miniaturized glucose monitors that can be read via a wrist band that is connected to the cloud. Hospitals can improve patient care via nonintrusive data collection, Big Data analytics, and intelligent systems.

The improvements to health care come through not just the medical care staff but the initiatives of medical equipment manufacturers to miniaturize and integrate their equipment with the goal of achieving more wearable, reliable, integrated, and effective monitoring and analysis equipment.

By making medical equipment smaller, multi-functional, and usable, efficiency is achievable through connecting intelligent devices to a patient's treatment plan in order to deliver medication to the patient through smart drug delivery systems, which is more accurate and reliable. Similarly, distributing intelligent devices over a network allows information to be shared among devices. This allows patient sensor data to be analyzed more intelligently, as well as monitored and processed quicker so that devices trigger an alarm only if there is collaborative data from other monitoring sensors that the patient's health is in danger.

Therefore, for the early adopters of the Industrial Internet, we can see that each has leveraged benefit in their own right, using innovation and analytics to solve unique problems of their particular industry.

The Why Behind the Buy

The IIoT has brought about a new strategy, which has arisen in industry, especially within manufacturing, and it is based on the producer focusing on what the customer actually wants rather than the product they buy. An example of this is why a customer would buy a commercial jet airliner. Is it because he wants one, or is it because he needs it to transport hundreds of his customers around the globe?

Traditionally, manufacturers set about producing the best cost-effective products they could to sell on the open market. Of course, this took them into conflict with other producers, which required them to find ways to add value to their products. This value-add could be based on quality, price, quantity, or perceived value for the money. However, these strategies rarely worked for long, as the competitor having a low barrier to entry simply followed successful differentiation tactics. For example, competitors could match quantity and up their lot size to match or do better. Worse, if the price was the differentiator, the competitor could lower their prices, which results in what is termed a race to the bottom.

Selling Light, Not Light Bulbs

What the customer ultimately wants the goods for is to provide a service (provide air transportation in the previous example), but it could also be to produce light in the case of a light bulb. This got manufacturers looking at the problem from a different perspective; what if instead of selling light bulbs, you sold light?

This out-of-the-box thinking produced what is known as the *outcome economy*, where manufacturers actually charged for the use of the product rather than the product itself. The manufacturer is selling the quantifiable use of the product. A more practical example is truck tires. A logistics company doesn't want to be buying tires for every truck in its fleet up front, not knowing how long they might last, so they are always looking for discounts and rebates. However, in the outcome economy, the logistic company only pays for the mileage and wear it uses on the tires, each month in arrears. This is a wonderful deal for them, but how does it work for the tire manufacturer? (We must stress a differentiator here—this is not rental.)

Well, it appears it works very well, due to the IIoT. This is feasible because each tire is fitted with an array of sensors to record miles and wear and tear and report this back via a wireless Internet link to the manufacturer. Each month the tire manufacturer invoices the logistics company for the wear of the tires. Both parties are happy, as they are getting what they originally wanted, just in an indirect way. Originally, the logistics company needed tires but was unwilling to pay anything over the minimum upfront as they assumed all the risk. However, now they get the product with less risk, as they pay in arrears and get the service they want. The tire manufacturer actually gets more for the tires, albeit spread over the lifetime of the tire, but they do also have additional services they can now potentially monetize. For example, the producer can supply data to the customer on how the vehicle was driven, by reporting on shock events recorded by the sensors or excessive speed. This service can help the customer, for example in the case of a logistics company to train their drivers to drive more economically, saving the company money on fuel bills.

Another example of the outcome economy is with Rolls Royce jet engines. In this example, a major airline does not buy jet engines; instead, it buys reliability from Rolls Royce's TotalCare. The customer pays fees to ensure reliable jet engines with no service or breakdowns. In return, Rolls Royce supplies the engines and accepts all the maintenance and support responsibilities. Again, in this scenario Rolls Royce uses thousands of sensors to monitor the engines every second of their working life, building up huge amounts of predictive data, so that it knows when a component's service is degrading. By collecting and storing all those vast quantities of data, Rolls Royce can create a "digital twin" of the physical engine. Both the digital and its physical twin are virtual clones so engineers don't have to open the engine to service components that are subsequently found to be fine, they know that already without touching or taking the engine out of service.

This concept of the "digital twin" is very important in manufacturing and in the Industrial Internet as it allows Big Data analytics to determine recommendations that can be tested on a virtual twin machine and then processed before being put into production.

The Digital and Human Workforce

Today, industrial environment robots are commonplace and are deployed to work tirelessly on mundane or particularly dirty, dangerous, or heavy-lifting tasks. Humans on the other hand are employed to do the cognitive, intricate, and delicate work that only the marvelous dexterity of a human hand can achieve. An example of this is in manufacturing, in a car assembly plant. Robots at one station lift heavy items into place while a human is involved in tasks like connecting the electrical wiring loom to all the electronics. Similarly, in smartphone manufacturing, humans do all the work, as placing all those delicate miniature components onto the printed circuit board requires precision handling and placement that only a human can do (at present).

However, researchers believe this will change in the next decade, as robots get more dexterous and intelligent. Indeed some researchers support a view of the future for industry in which humans have not been replaced by robots but humans working with robots.

The logic is sound, in that humans and robots complement each other in the workplace. Humans have cognitive skills and are capable of precision handling and delicate maneuverings of tiny items or performing skills that require dexterity and a sense of touch. Robots on the other hand are great at doing repeatable tasks ad nauseam but with tremendous speed, strength, reliability, and efficiency. The problem is that industrial robots are not something you want to stand too close to. Indeed most are equipped with sensors to detect the presence of humans and to slow down or even pause what they are doing for the sake of safety.

However, the future will bring another class of robot, which will be able to work alongside humans in harmony and most importantly safely. And perhaps that is not so far-fetched when we consider the augmented reality solutions that are already in place today, which looked like science fiction only a few years ago.

The future will be robots and humans working side by side going by the latest research in IIoT. For example, robots are microcosms of the Industrial Internet, in so much as they have three qualities—sensing, processing data, and acting. Therefore, robots—basically machines that are programmable to replace human labor—are a perfect technological match for the IIoT. Consequently, as sensor technology advances and software improves, robots will become more intelligent and should be able to understand the world around them. After all, that is not so far away as we already have autonomous cars and drones. Expect robots to be appearing in supermarkets and malls near you soon.

References

https://www.ge.com/digital/industrial-internet

https://www.accenture.com/us-en/labs-insight-industrial-internet-of-things.aspx

http://www.forbes.com/sites/bernardmarr/2015/06/01/how-big-data-drives-success-at-rolls-royce/#206a7e173ac0

http://zwcblog.org/2014/08/08/selling-light-not-light-bulbs/

http://news.mit.edu/2015/giving-robots-more-nimble-grasp-0804

CHAPTER 2

Industrial Internet Use-Cases

The potential for the Industrial Internet is vast with opportunities spread over wide areas of productivity, such as logistics, aviation, transportation, healthcare, energy production, oil and gas production, and manufacturing. As a result, many use-cases will make industry executives wake up and consider the possibilities of the IIoT. After all, industry only requires a minimal shift in productivity to deliver huge revenue, an example is that even an increase of 1% of productivity can produce huge revenue benefits such as aviation fuel savings. In order to realize these potential profits, industry has to adopt and adjust to the Industrial Internet of Things.

However, spotting, identifying, and then strategically targeting the opportunities of the IIoT is not quite as easy as it might seem. It is important, therefore, to create use-cases that are appropriate to vertical businesses. For instance, the requirements of manufacturing differ from logistics, which also differs to healthcare. Similarly, the innovation, expertise, and financial budget available to deliver specific industry applications will have many diverse constraints. For example, healthcare will consume vast amounts of expenditure with little or no financial return; in contrast, the oil and gas industry will also require immense operational and capital cost but will likely deliver huge profits. Similarly, logistics—which is very reliant on supply chain, product tracking, and transportation—will have different operational requirements. However, what the IIoT offers is a potential solution for all vertical industries, by utilizing the advances in sensor technology, wireless communications, networking, cloud computing, and Big Data analysis. Businesses can, regardless of their size and discipline, leverage these technologies in order to reap the rewards of the IIoT.

To illustrate the potential benefits and advantages to individual industrial disciplines, consider the following use-cases.

Healthcare

In this example, we will strive to explain how utilizing IIoT technology can unlock and deliver value to the heath care industry. In healthcare, and it wasn't so long ago, doctors made house visits to those too sick or incapacitated through injury to make their way to the doctor's office. However, this was time consuming and costly. Consequently, doctors restricted home visits to only those who in the doctor's experience deemed sufficiently seriously incapacitated through illness or injury, and everyone else had to turn up and take their place in the doctor's office queue. This policy, though understandable, was seriously inconvenient for both patients and doctors, especially patients in rural areas who might have to drive considerable distances while suffering from the debilitating effects of illness or physical injury. Therefore, an alternative arrangement was always desirable.

That is why Guy's and St. Thomas's Nation Health Service Foundation Trust in the UK are piloting the use of smartphones to use as health monitors. The patient's kit compromises a smartphone, scales, blood oxygen sensors, and a blood pressure cuff. The idea is that the patients will take daily readings of their weight, heart rate, blood pressure, and oxygen levels, then upload the data to the smartphone via Bluetooth to be sent to BT's telehealth service.

Nurses at the service then analyze the data. If there are any abnormalities in the data, the nurses will discuss issues with the patients. By using these homecare kits, patients have more control over their own condition and can manage their own chronic medical conditions in their own homes. It is hoped that the pilot project, which is being tested on 50 heart failure patients, will ultimately save lives.

Another example, of a state-of-the-art IIoT project in today's healthcare environment is the initiative adopted by Scottish health chiefs to provide a means of automation, supervision, and communication for remote outpatients.

The robot—known as the Giraff—is being used in the homes of patients, particularly those suffering from dementia in the Western Isles and Shetland to allow them to continue living independently. The robots are designed to provide reassurance to friends and family, by enabling a relative or carer to call up the Giraff from a remote computer or smartphone from any location. The 3G audio/video channel displays the carer's face on the Giraff's video screen, allowing them to chat to the patient via a Skype-like video call.

The Giraff launched in 2013 as a pilot trial. The Giraff robots are just under five feet tall with wheels, and a video screen instead of a head. They are fitted with high-definition cameras to monitor the home and provide remote surveillance. The Giraff allows relatives and carers to keep a vigilant eye on the patients, to ensure they are taking their medication and eating meals, while also providing a method for social exchange potentially from hundreds of miles away. The carer can also manipulate the robot and drive the robot around the house to check for any health or safety issues.

The use of assistive technology is sometimes targeted at specific patients, and, as such, the Giraff would have a specific rather than a generic application. It was initially feared that older patients suffering from dementia would react badly to the presence of a robot. On the contrary, it appears that they found the robot good company, even though it could not hold a conversation (although the likes of Siri could address that immediate problem and neither can a dog or cat). Furthermore, earlier trials in Australia showed that people with dementia were not afraid of the machines. They hope the robots will help people living alone in remote areas to feel less lonely.

Another personal healthcare robot is Baymax, which is a robot with a soft synthetic skin that can detect medical conditions (this was an initiative based on a fictional Disney character in *Big Hero 6* but it may not be far from becoming reality). Early versions of a robot teddy bear, developed by MIT Media Lab, are now being put through their paces in a children's hospital in the United States. An updated version of the bear has been fitted with pressure sensors on two of its paws and several touch sensors throughout its body parts. The screen of the smartphone device in the robot's head shows animated eyes. The robot can use the phone's internal speaker, microphone, and camera for sensing changes in a child's well-being.

Oil and Gas Industry

The Oil and Gas industry depends on the development of high technology as well as scientific intelligence in the quest for discovery of new reservoirs. The exploration and development of newly discovered oil and gas resources requires modern sensors, analytics, and feedback control systems that have enhanced connectivity, monitoring, control, and automation processes.

Furthermore, the oil and gas industry obtains for process vast quantities of data with relation to the status of drilling tools and the condition of machinery and processes across an entire field-installation.

Previously, technology targeted oil and gas production but geologists had limited ability to process the vast amounts of data produced by a drilling rig, as there was just so much of it and storage was expensive and just not feasible. Indeed, such was the vast amount of data collected, up to 90% would be discarded, as there was nowhere to store the data let alone have the computational power to analyze it in a timely manner.

However, the Industrial Internet of Things, (IIoT) has changed that wasteful practice and now drilling rigs and research stations can send back the vast quantities of raw data retrieved from drilling and production sensors for storage and subsequent analysis in the cloud. For example, drilling and exploration used to be expensive and unpredictable as it was based on geologist's analysis of the mapping of the sea floor. This proved to be unpredictable and, as a result, major oil and gas exploration and producers are transforming their infrastructures to take advantage of the new technologies that drive the Industrial Internet. These technological advances, such as high bandwidth communications, wireless sensor technology, cloud data storage with advanced analytical tools, and advanced intelligent networks are enabling systems that enhance the predictability of field research, make research more predictable, reduce exploration costs, and also eventually lower field operation expenses.

New industry regulations for well monitoring and reservoir management have, on top of other technical demands, pushed field operators to find efficient ways of addressing existing operational constraints. For example, in the 1990s and 2000s, it was commonplace for field operators to dump almost all of the data they collected during drilling due to a lack of processing and communication capabilities; the amount of data was just too vast to accommodate. In mitigation, most of the data was only relevant to the time it was generated—for example, the temperature of the drill bit, or the revolutions per second—so was only useful at that specific time.

However, with the advances in technology, specifically in down-hole sensors and the subsequent massive influx of data from down-hole drilling tools, which required advanced analysis in both real-time data streaming as well as historical and predictive analysis, demands for more innovative solutions have increased.

Fortunately, just as the demand has grown for such vast data analytics within the oil and gas industry, another technology has come to the fore that provides the necessary compute, data storage, and the industrial scalability to deliver real-time data analysis. Additionally cloud technology is able of batch processing Big Data mining, for historical understanding and predictive forecasting.

Cloud computing and the Industrial Internet now provide the technology to make gathering, storing, and analyzing vast quantities of data economically feasible.

However, the advent of the Industrial Internet has delivered far more than economic and scalable cloud services in compute, storage, and data analytics; it has changed industry profoundly. For example, industry now has the ability through interconnectivity to connect intelligent objects—machines, devices, sensors, actuators, and even people—into collaborating networks, an Internet of Things. At the same time, the designers of these intelligent, smart things have built in self-diagnosis and self-configuration, which greatly enhances reliability and usability. In addition, device connectivity, the requirement for cables and power, which was once a real problem, has been alleviated by wireless communication. New wireless technologies and protocols, along with low power technologies and component miniaturization, enable sensors to be located anywhere, regardless of size, inaccessibility, or cabling restrictions.

Connectivity is at the core of the Industrial Internet; after all, it requires communications over the Internet and interaction with the cloud. Therefore, the communication protocols are all important and this has produced new protocols such as 6LoWLAN and CoAP, which we will discuss in subsequent chapters at a technical level later. These may work well for some industrial use-cases that have low capability devices deployed in end-to-end connectivity.

However, for all systems there are only two ways to detect a remote node's status—the sensor sends data back to the controller, for example as an event or the controller polls the node at programmable intervals to obtain the nodes status. Both of these are inefficient, but there is a better way (discussed in detail later), which is the publish/subscribe software pattern. It's a preferable technique as it can instantly inform a subscriber across a common software bus of a change if that subscriber has noted an interest. This is preferable to the subscriber polling the publisher for any updates, as it is far more efficient and quicker. However, not all publish/subscribe models work in the same manner. MQPP and XMPP are very popular as they are well supported; however, they do not support real-time operations, so are not well suited to industrial scenarios.

The data distribution system does support real time operation and it is capable of delivering data at physical speeds to thousands of recipients, simultaneously, with strict control on timing, reliability, and OS translation. These are hugely important qualities when deployed in an industrial environment, such as the oil and gas industry.

It is these new IoT protocols and technologies that have provided the means to change oil and gas exploration and field production beyond what was previously feasible.

As an example of how the oil and gas industry utilizes DDS as a publish/subscribe protocol, let's examine how they have integrated it into their operational processes.

The first example shows how IoT has enabled remote operations of drilling rigs by automation. This is not only cost effective at a time when field experts are becoming a rarity, but also beneficial with regard to field efficiency, safety, and well quality. It can also lead to—via advanced sensor technology being self diagnostic and self-configurable—a significant decrease in downtime and equipment failures.

Figure 2-1 shows a block illustration of an automated remote control topology, whereby a high-speed DDS data bus connects all the sensors and actuators with a process controller, which automates the process of drilling and completion.

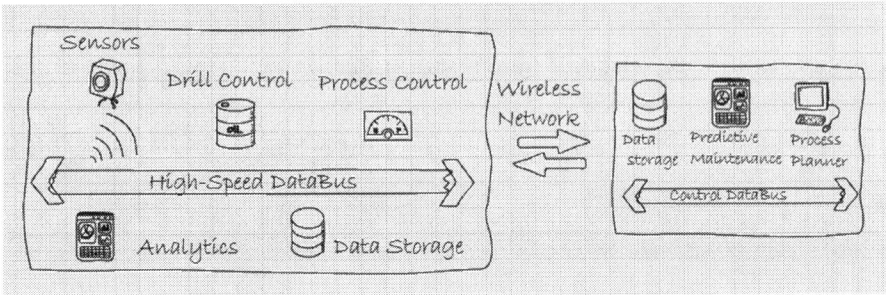

Figure 2-1. DDS data bus

In addition to automation, the design also facilitates the remote collection and analysis of operational data, equipment health, process activity, and real-time streaming of equipment log data.

The high-speed connectivity provided by either wireless or fiber optic cables connects the field well with the remote control station and ultimately with the enterprise systems. Data collected from the field station, via the DDS bus, can be stored for future historical and predictive analysis. This will allow on-shore analysts and process planners to adjust and control the well operations by sending corrective feedback to the well systems.

Another opportunity that the IIoT delivers is that of enabling massive data collection and subsequent analysis. Prior to the advances and public access to the vast resources in cloud computing, it just was not feasible or economical for even cash rich oil and gas companies to hoard vast quantities of data. After all, the amount of data generated by an operational drilling or production oil well can be vast. However, now that has changed with the Industrial Internet technologies being able to accommodate both the storage and the compute power to analyze these vast data sets. A typical use for such technology would be in intelligent well monitoring, whereby entire fields of sensors are monitored and the data accumulated to provide data to a remote control center for historical and predictive analysis.

Furthermore, an additional use-case for the oil and gas industry of IIoT is in the deployment of intelligent real-time reservoir management. In order to benefit from analytics, whether they are historical or predictive, all the systems within the ecosystem must be connected and contribute to the pool of data. The larger the pool of data, the more reliable the results of algorithms will be, as it can mitigate the risk of irregular data patterns that do not necessarily reflect the true nature of the process. For a simplistic example, consider tossing a coin ten times and then ten million times when considering the probability of heads or tails. This, connectivity of systems is even more important when dealing with real-time analytics on streaming data, where real-time analysis and feedback is required. However, the topology of large-scale analytical networks is not trivial, with systems interfaced and data driven via a data bus to the cloud or to streaming analytical tools. With DDS, a designer can decouple the complexity of the physical connections among computers, machines, systems, and sites by provision of a single logical data bus.

Finally, a last use-case example shows how deploying IIoT protocols and technology can ease the production and deployment of industrial platforms as it decouples software from the operating system, thereby making application development more agile, quicker, and cheaper.

The real potential of the IIoT is to create new, intelligent ways of working, through automation, intelligent machines, and advanced analytics. In the oil and gas industry, IIoT methods and technologies are already being adopted to reduce costs and increase efficiency, safety, and ultimately profits. However, the future of the IIoT must integrate with the cloud, which then has the potential to merge local applications into larger regional or global systems, to become a network of systems that deliver the full potential of Big Data analytics to industry.

Smart Office

Buildings are critical systems, and they are responsible for approximately 40% of the total EU energy consumption. What is worse is that buildings are also to blame for 36% of green house gas emissions. However, controlling or reducing these figures is not easy. Even with a three-pronged strategy, such as improving building insulation and energy efficiency and providing better building control systems, progress has been painfully slow. Typically, this is due to the results of several conditions. The first of these strategies—improving insulation—is a major cost saving incentive for any building as it reduced heating or cooling costs to the inhabitants. Furthermore, it reduces energy costs and reduces CO_2 emissions and is easy to implement into the design and installation of new buildings, but very expensive and difficult to deploy into existing buildings. The reason for this is that most older buildings were simply not designed to be energy efficient.

The second strategy for improving the building's energy efficiency, for example, by changing light bulbs and strip lighting for LED lights, is gaining some traction but is still under exploited. This may be due to a failure to get the message across to property owners and businesses. However, the third strategy, improving building management through automation control systems, can provide the potential to improve building energy efficiency and reduce green house emissions.

Unfortunately, like installing insulation into existing buildings, especially older ones, deploying a building control management system is a painful task, both in financial costs and in business disruption. Previously, installing sensors and actuators (such as on radiators or on AC units) required major refit work. However, with the recent advances in technology and the IoT in particular, sensors and actuators are now "smart" and can use wireless communications, which greatly reduces the disruption and much of the cost.

The integration and development of sensors, devices, and protocols based on the IoT are important enablers of applications, for both industries and the general population, by helping to make smart buildings a reality. IoT technology allows for the interaction between smart things and the real world, providing a method for harvesting data from the analogue world and producing information and knowledge in the digital world.

For example, a smartphone has built-in sensing and communication capabilities, such as sensors for acceleration, location, along with communication protocols that support Wi-Fi, SMS, and cellular. They also have NFC (near field communication) and RFID (radio frequency identification), both of which can be used for identification. Consequently, the smartphone provides the means to capture data and communicate information. Also, the ubiquity and user acceptance of the smartphone makes them an ideal HMI (human machine interface) for smart buildings, where users need to control their own environmental conditions.

Nevertheless, the IoT comes with its own set of problems, such as the management of huge amount of data provided in real time by a large number of IoT devices deployed throughout the building. Additionally, there is the problem related to the interoperability of devices, and furthermore the integration of many proprietary protocols and communication standards that coexist in the marketplace. The protocols that are applicable to buildings (such as heating, cooling, and air conditioning machines) may not be available on devices presently available off-the-shelf. This needs addressing before wide-scale adoption is achievable.

One of the main problems with installing traditional building management systems (BMS) into existing and especially older buildings is that the traditional methods are often based on specialized protocols, which we will discuss later, such as BACnet, KNX, and LON. In addition, the alternative WSN (wireless sensor networks) solutions are based on specific protocol stacks typically used in building control systems, such as ZigBee, Z-Wave, or EnOcean. The deployment is much easier than with the BACnet wired bus, but they still have issues with integration into other systems.

To this end, in 2014, IoT6 (a European Union working group) set up a testbed for a smart office to research the potential of IPv6 and related standards in support of a conceptual IIoT design. The aims were to research and test IPv6 to see whether it could alleviate many of the interconnectivity and fragmentation that currently bedevils IoT implementation projects. The methods the IOT6 group decided on was to build a test office using standard off-the-shelf sensors, devices, and protocols. IPv6 was preferable but not always an option due to lack of availability. The devices were connected via a service-orientated architecture (SOA) to provide Internet services, interoperability, cloud integration, mobility, and intelligence distribution.

The original concept of the IOT6 Smart Office was to investigate the potential of IPv6 as a common protocol, which could provide the necessary integration required between people and information services, including the Internet and cloud-based services, the building, and the building systems.

The IOT6 team hoped to demonstrate that by better control of traditional building automation techniques, they could reduce energy consumption by at least 25%. In addition, they hoped to ease the deployment and integration of building automation systems, something that is typically costly and requires refits and expensive installation. They also looked to improve the management of access control and security by utilizing smartphones as an HMI.

With regard to the integration of people and the building information services, the testbed would provide a location, a smart office that was fully equipped and operational. It would provide a meeting and conference rooms, and they would also provide for innovative interfaces within the building (virtual assistant, etc.) that would enable users to interface with their environment and customize the actions of sensors controlling things like the temperature, lights, and blinds. Furthermore, the office would have full information and services, such as computers for Internet access and displays to provide real-time information on the state of the world. In addition, the smart office would provide a professional coffee machine—a machine that provides hot water 24/7.

Chapter 2 | Industrial Internet Use-Cases

One of the goals of the IOT6 testbed was to provide a platform for testing and validating the interoperability among the various of-the-shelf sensors and protocols and the conceptual architecture of the Industrial Internet of Things. They were determined to interconnect and test wherever possible multi-protocol interoperability with real devices through all the possible different couplings of protocols (among the selected standards). Also, they wanted to test and demonstrate various innovative Internet-based application scenarios related to the Internet of Things, including business processes related scenarios. In addition, they planned to test and demonstrate the potential of the multi-protocol card, IPv6 proxy's for non-IP devices, and estimate the potential scalability of the system. Furthermore, they would deploy and validate the system in a real testbed environment with real end users in order to test the various scenarios.

The four scenarios tested were:

- The first scenario involved the building maintenance process, which is the process of integrating IPv6 with standard IoT building control devices, mobile phones, cloud services, and building management applications.

- The second scenario addressed user comfort in the smart office and this is really where the office does become intelligent or "smart". In this scenario, a user is identified by his mobile phones, NFC, or RFID, and the control management system will adjust the environment to the user's pre-set or machine learned preferences, such as temperature or light levels that provide the user with a welcoming ambience. When a visitor arrives, detected again by RFID on their mobile phone, the CMS can turn on the lights in the reception area and play music and video, again to provide a welcoming atmosphere. When the last person leaves the smart office, detected by presence detectors, the CMS will turn off the lights and reduce the HVAC to the standby condition.

- The third scenario related to energy saving and awareness. In this scenario, the intention was to demonstrate the use of IPv6, with a focus on energy management and user awareness. The intention was to allow a user, when entering an office, to adjust the environment using their mobile phone app. The mobile app will display current settings and when the user selects to change the setting the mobile app will display the energy consumption implications of such modifications. Once the user leaves the room, the system returns the settings to the most economical energy configuration.

- The fourth scenario entailed safety and security and focused on intrusion detection and fire-detection. In this scenario, the system learns of a security issue due to presence detectors, which notify the system of someone being in a room that is supposedly empty, or magnetic switches on windows or doors trigger the alarm. Similarly, temperature sensors or smoke detectors can trigger fire-detectors. In both cases, the system looks up the IP addresses of the closest security server and possible backups. The system contacts the local data server by sending the data by anycast with QoS and priority routing. If it does not receive a reply, it sends duplicate data to another group of security servers. The system also contacts the closest duty security agent, who can then access the location via remote video using their mobile phone app.

The IOT6 group discovered through their technical analysis of the Smart Office that there were many significant improvements when deploying a building control management system using IoT devices based on an IPv6-aware protocols such as 6LoWPAN and CoAP on a native IPv6 network (discusses later in the technical chapters). They reported improvements in ease of deployment, scalability, flexibility/modularity, security, reliability, and the total cost of deployment. The technical reports key performance indicators focused on energy savings and improvements in energy efficiency.

Logistics and the Industrial Internet

Logistics has always been at the forefront of the IIoT, as so much of the opportunities and techniques are a perfect match for the logistics industry. Therefore, it is no surprise that the logistics industry has been using many of the sensors and related technologies associated with the IIoT for years. For example, logistics have been using barcode technology in packaging, pallets, and containers for many years as a way to monitor inbound deliveries and outgoing dispatches from warehouses. This was a huge advance from the previous method of opening each attached delivery note and physically checking the items. However, using manual barcode scanners was still labor intensive and although accurate if performed diligently there were still pallets overlooked or products going undetected. In order to address these inventory control processes, logistic companies sought an automated solution using IIoT techniques and wireless technologies.

The solution is to use embedded RFID tags and the associated RFID readers, which can scan entire rows or stacks of pallets queued at the inbound gate simultaneously. This is something a barcode reader had to perform one at a time, which is an improvement in speed and accuracy as every RFID tag in radio range on every pallet, whether visible or not, is read by the system. The RFID reader automatically records the RFID tag's information such as the order ID, the manufacturer, product model, type, and quantity, as well as the condition of the items before automatically recording the delivery in the ERP system.

Once the inbound delivery has been recorded and the items moved to the correct stock location, the tags can be updated to show the relevant stock details, such as part numbers and location. They can also communicate other information using temperature and humidity sensors and send information regarding the environmental storage conditions. This allows warehouse staff to take action before the stock becomes damaged.

Another major benefit of using RFID tags is that they allow for fast and accurate audits of stock. Stock level is managed through an ERP application interfacing with the RFID readers, so changes in stock levels are updated automatically and stock levels are continuously updated and discrepancies are immediately alerted.

Similarly, for outgoing stock control when an order is dispatched, an RFID tag reader can read all of the pallet tags as they pass through the outbound gates and automatically adjust the stock holding for every item simultaneously, while also updating each order's ERP delivery ticket as being complete and in good condition.

Automating stock control task such as delivery and dispatch has improved operation efficiency and stock control accuracy because booking in and out products to warehouses is now a trivial process.

Such is the competitive advantage gained by adopting sensor technologies in improved operational efficiency, for example faster, accurate, and cost-effective warehouse management, logistic companies are always keen to explore new IIoT initiatives. The areas that are proving appetizing to logistic companies are with optimized asset utilization, whereby a centralized system can monitor the condition, status, and utilization of machinery and vehicles. This is important for warehouse managers as they often unintentionally overutilize some assets while underutilizing others. For example, a forklift truck may sit idle in another area of the warehouse, when other forklifts and drivers are working continuously.

Another operational issue is that in large warehouses, forklift productivity can be problematic. This issue arises as the result of drivers needing to find the stock locations and navigate the aisles and rows trying to locate the correct products. Using a combination of location sensors, barcodes, RFID tags, and ERP stock data, it is possible to instruct the driver to the location of the stock items and provide directions of how to get there from the driver's current location. This is a method adopted by Swisslog's SmartLIFT technology, which uses directional barcodes on the ceilings and aisles, in addition to forklift sensors and warehouse stock location data to create a visualization of each stock location in relation to the forklift's current position. By working similar to GPS, the system informs the driver as to the best route to the stock. SmartLIFT technology improves forklift utilization and reduces stock-handling errors considerably.

Forklifts are the cause of over 100,000 accidents each year in the United States alone, with almost 80% involving pedestrians. Therefore, the logistics industry is keen to utilize IIoT to prevent many of these accidents. There are several ways that IIoT can help, for example, by using sensors, cameras and radar on forklifts to warn the driver of the presence of pedestrians and another forklift. Ideally, a forklift would communicate with other forklifts, ensuring they were aware of one another to take avoiding action, such as slowing or stopping at blind intersections if another forklift is detected in the immediate vicinity.

However, in the developed world it is still far more common to pick-by-paper, which is the term applied to the manual human picking of goods from a shelf. Forklifts, autonomous vehicles, and robots are great for heavy lifting of large pallets, but not much use for picking small intricate articles out of a stock bin. This is where human workers are in their element. Remember all those pedestrians being injured in the warehouse by forklifts? Well those pedestrians are most likely to be the pick-by-paper workforce. These are workers employed to collect individual stock items from a list. It is not very efficient and they have the same problems as the forklift drivers, finding their way around the warehouse and locating the stock.

However, help is at hand through augmented reality. The most commonly known augmented reality device is Google Glass; however, other manufacturers produce products with AR capabilities. Where augmented reality or, for the sake of explanation, Google Glass, comes into logistics is that it is extremely beneficial for human stock pickers. Google Glass can show on the heads up and hand free display the pick list, but can also show additional information such as location of the item and give directions on how to get there. Furthermore, it can capture an image of the item to verify it is the correct stock item. Where items are practically identical to the eye, for example a computer chip, or integrated circuit, hands-free, automatic barcode scan ensures correct item identification. Furthermore, augmented reality accelerates training, and since the stock pickers are often seasonal temporary workers, this is very important. The technology also allows for hands-free use, which leads to greater productivity, as workers can find the items far more quickly, which greatly increases efficiency while eliminating pick errors.

Augmented reality glasses are similarly suited to freight loading whereby forklift drivers can do away with the fright load sheet, which tells them the order each pallet has to be loaded onto the truck. In the same manner as with the stock picker, the forklift driver will see displayed on the glasses the relevant information, which increases load times as the driver has hands-free information so does not have to keep stopping to refer to a printed list.

Another very promising use-case for IoT and augmented reality is using document scanning and verification. In its most simple use-case delivery drivers can check that a load is complete with every package or pallet accounted for and loaded. In a more advanced use-case, the glasses could be used to scan foreign documentation, the type used in international trade for compliance with import export laws. The augmented reality device's IoT integration could enable a driver to scan the document while the software looked for keywords, phrases, and codes required for the document to be valid. This could save many wasted hours at ports and borders correcting incomplete or inaccurate paperwork.

The potential IIoT use-case for logistics goes beyond the warehouse and has interesting applications in freight transportation. Currently, logistics companies perform track and trace and they can monitor the location of pallets on an aircraft mid-flight or a container on a ship in the middle of the ocean. Despite these capabilities, the industry is looking forward to a new generation of track and trace, which would bring improvements in speed, safety, accuracy, and security. For example, theft of freight goods is still a major issue and with more robust IIoT solutions deployed it would enable freight goods to be tracked meter by meter from dispatch to arrival. Advanced telemetric sensors in trucks and RFID tags on goods will allow for accurate and predictive location and condition monitoring. Multiple sensors in freight goods will monitor conditions such as temperature, humidity, shock, or even if a package has been opened, which might indicate potential theft.

The trucks themselves can use advanced telemetric sensors to predict when and how the vehicle will require maintenance and to automatically alert the driver and maintenance crews and even schedule a window for the required service. However, it is not just the trucks that require monitoring; drivers have to work long hours, sometimes in hazardous conditions, and fatigue can be a health and safety issue for themselves and other road users. There are already technologies in use that help detect driver fatigue. For example, Caterpillar uses infrared cameras to monitor the driver's eyes, and a computer monitors blink rate and pupil size. Should it detect the drivers are sleepy, it will alert them using audio alarms and seat vibrations.

Another possible use-case is in supply chain management where the predictive analysis techniques of Big Data can come into play. The world's largest logistic companies need to know the latest current events on a global scale, such as the political climate as well as the local weather conditions that affect traditional trade routes. They need to know of impending strike action by traffic controllers or crane drivers in a shipping port, as these could cause massive disruption and have a knock-on effect to a customer's stock inventory levels. However, with trucks and goods bristling with sensors, it is now possible to harvest this data on a global level. When combined with data on current affairs, natural disaster, socioeconomic unrest, and similar threats to traditional trade lanes. It will be possible to manage threats proactively by moving cargo from air to sea or vice versa to mitigate the threats of strike action.

Similarly, urgent cargo routes can be altered in real-time if predicative analysis of all the global data shows a high risk of civil unrest or bad weather on route, which could seriously delay delivery. Predictive analysis through Big Data is becoming a required tool for business intelligence analysis; it is believed that over 80% of businesses will adopt it in one form or another in 2016. The promise that predictive analysis holds for global logistics is that they will be able to take proactive action to mitigate potential threats to their operations and keep their freight moving to the destination.

Retail

Retailers, like most businesses, suffer IT costs and overheads, which directly affects their profits, as they must pass these costs onto the customer as part of the cost of goods. Therefore, it is in the retailers' interest to reduce, manage, or at least leverage value out of these IT costs.

IT costs incurred by a retailer typically come in the form of IT systems required to run the business and the follow-on costs of supporting those systems. In order to reduce their IT costs, the retailer must find a way to meet their requirements using a cheaper, more efficient technology. The high IT costs in retail are generated by replication of hardware and software licenses, IT support such as site maintenance, and break-fix callouts. Any solution that reduces those variable high overheads and transforms them into a lower fixed cost is a favorable alternative.

Retailers require a means of collecting payments, temporarily securing cash and credit receipts, recording sales transactions, refunds, partial payments, the running cash balance, and importantly produce an end-of-day sales report. All this information is required for the retailer to reconcile the end-of-day sales reports with the cash in hand. This might not seem much but for a retailer with say 1,000 stores reconciling the total sales at the end of day, both individually and collectively can become an onerous task.

Traditionally, the retailer accomplished end-of-day sales reconciliation by using cash registers overseen by store managers. Cash registers evolved into point of sales (POS) machines, based on PC technology with specialist peripherals, such as a cash box and a barcode scanner. Therefore, as this is at the technology level simply a computer with POS software replicated in every store, we can optimize this by shifting the intelligence and the majority of the IT burden from the individual store into a central location—the cloud.

By shifting the IT burden out of the stores to a networked environment, the head office, and into the cloud, we greatly reduce the capital (Capex) and operating (Opex) costs. We reduce equipment costs for the stores as now the POS device simply needs to run a web browser as the POS application runs in the cloud. Therefore, sales assistants can use mobile devices such as tablets or smartphones, freeing them from the fixed sales point, and this can reduce the burden of staffing a busy store. These savings alone can turn a borderline store into a viable one.

However, running the POS application has greater operational value, such as central administration and management. Previously with legacy POS solutions each store completed its own end-of-day report and sent it to the retail head office, which had to collate all the reports into one unified report. This could take until late the next day before the manager/owner had a clear picture of the overall sales performance. However, with a cloud POS application, the individual and collective reports are available at any time and from anywhere—and they are accurate.

Having an immediate view of the individual and overall sales performance optimizes other areas of the business, such as stock control (inventory). After all, every sale requires a stock replenishment or at least set the process in motion. By storing all the individual stores data in a central multi-tenant database, which segregates each stores sales and stock, the cloud application can automatically handle sales and stock management both at an individual and collective perspective, and do it in real time, not late the next day.

Central management is a very important feature of a cloud POS as it allows management to apply a central policy to every store and POS device with the click of a button. It also enables the automation of processes, procedures, and workflows, which manage stock replenishment.

Real-time reporting and visibility is also a major cloud service that shortens the decision-making process. Having immediate access to reports and live status reports enables greater control and more timely management. For example, during sales promotions the performance at individual stores are tracked as they happen, tactics altered, and decisions made to immediately replenish dwindling stock, as events unfold and not on yesterday's stale data.

As we have seen, retailers can reduce costs by taking a cloud POS strategy to IT and the remaining costs managed and leveraged to provide value to the business. Direct IT cost savings come about through reduced hardware/software and support/maintenance. A cloud POS solution removes these costs replacing them with one fixed monthly fee. This alone reduces the cost of goods and increases profit. Additionally, cloud POS provides the services to create value through efficiency and control, and the ability to manage the business in real time.

IOT Innovations in Retail

Innovations in the retail industry have gone well beyond just operational efficiency; the retail industry has positively embraced the IIoT. Here are some examples of the change in mindset.

The IIoT will provide a means for innovation for retailers previously unimaginable, for example, it facilitates, a means to communicate directly to customers through omni-channels, such as in-store interactive advertising, web promotions, social media, video, and augmented reality. Let's not leave the traditional media, such as newspaper, magazines, and TV/radio behind, as they still have massive consumer reach. However, in order to stay relevant, retailers have to deliver the right message via the popular channels that focus on their customers' preferences, and importantly look forward to getting their message across to the next generation of customer.

This approach requires a transformation in how retailers adopt the IoT as it is at the heart of this transformation. The potential of the omni-channel reach of the IoT is that it connects people, machines, items, and services, well beyond any single communication channel, such as TV. Therefore, retailers, in order to streamline the flow of information, must ensure that their advertising strategies are delivered over omni-channels. It's also important that retailers target customer real-time decisions and heighten consumer experiences, as this is paramount in the Internet age.

Leading fashionable retailers that target young fashion-conscious consumers are already investing billions in the IoT—think about mobile phone shops and high street fashion stores. Their adoption of high-tech advertising and providing in-store virtual customer experience provided by the adoption of the IoT has realized the financial returns beyond expectations. The operational and financial benefit that retailers have reaped range from efficient inventory management to real-time customer targeted promotions, and in-store entertainment that increase footfall, develops the brand and ultimately increase sales. These pioneers of IoT have identified early that it was necessary to evolve and transform their business practices to conform with the shift in their customers' lifestyles. They also identified that, in time, the IoT will touch nearly every area of retail operations and customer engagement.

Retailers that have managed to successfully navigate this transformation, from traditional to state-of-the-art, have not only embraced innovation but have realized the potential of IIoT.

Retailers may well understand the concepts of the IIoT, as well as the business and marketing potential. Unfortunately, they find there is rarely concrete evidence of the specific competitive advantage. More often than not, the IIoT is portrayed as deploying high technology, such as augmented reality, RFID customer tracking with personalized advertising, and similar marketing concepts that would not fit easily with most retailers' current or even future customers.

The advantages of IoT are not delivered just through enhanced customer experience—many of the benefits come in the back store, in stock control, inventory management, perishable and cold chain management, and for larger operations, digital signage, fleet management, and smart fulfillment centers. As an example, three of the largest supermarkets in the UK reported savings of 50% after the adoption of the IIoT.

For some retailers deploying IIoT solutions, it has meant installing a vast range of bewildering technologies, including hardware, sensors, devices, apps, telematics, data, and connectivity to the cloud. Even so, the operational benefits are clear, and in addition, there are further gains that can be reaped from interactive advertising. This innovative form of advertising works by embedding NFC or RFID tags to products, which provide customers with additional information when they are in close proximity. An example, of this is when a potential customer stops in front of a product on a supermarket shelf, a digital sign located below the product is activated to provide further information, such as nutritional data or in the case of clothing, social-media ratings and perhaps a current discount. These may sway the customer's purchasing decision.

The trouble is though, is that all of this can be a big turnoff to the customer. Not everyone wants to have such intrusive marketing thrust upon them. As a result, retailers are attempting to design the customer experience, which they now realize can become as important as advertising the products. With the shopping experience becoming increasingly more valued than material items to the customer, it makes sense to begin planning for new products and services with the customer's experience in mind.

What retailers that operate in a business-to-consumer or a business-to-business model have realized is that adopting IoT concept and technologies will simplify life for the target audience and offer a fulfilling customer experience. The intent is that the IoT initiatives and technologies deployed will not just push products, marketing, and services, but will contribute to the overall enhanced customer experience, which results in higher individual sales and greater gross profits.

Of course, not all customers are the same. Some absolutely revel in high technology as can be seen though the success of stores deploying augmented reality. In these stores, retailers have gone a step beyond inventory control and NFC card payment retailers and have provided a virtual magic mirror.

Augmented reality is a new trend in retail as it provides a way for customers to evaluate products interactively and compare them to other similar products or consider their suitability to the environment they would be situated. Examples of augmented reality are the latest IKEA catalogue, a mobile app that enables customers to virtually simulate having the items of furniture in their real living room. The customer can arrange the virtual furniture in different location checking for dimensions, color schemes, and alter their choices to suit.

Similarly, an LA-based fashion company, Uniquo's magic mirror, has created a stir, as it allows customers to try on one item of clothing, then the AV mirror recognizing the size, style, and product can change colors to show available alternatives, without the customer having to change each time.

However, IoT and augmented reality do not stop there. Virtual interfaces can go even further as demonstrated by Yihaodian, the largest food e-retailer in China. The company recently announced that it was going to open the first AR supermarket chain in the world. Each of these virtual supermarkets has a completely empty floor space and situated near high footfall areas (e.g., train or subway stations, parks, and universities). The interesting thing is that while the naked eye will just see empty floors and walls, people using an AR-capable device, for example Google Glass, will see shelves filled with vegetables, fruit, meat, fish, beer, and all sorts of real-world products. To buy these virtual products, the customer scans each virtual product with their own mobile devices, adding it to their online shopping carts. They subsequently receive delivery of the products to their homes.

References

http://www.giraffplus.eu/

https://www.rti.com/whitepapers/5_Ways_Oil_Gas.pdf

http://www.dhl.com/en/about_us/logistics_insights/dhl_trend_research/Internet_of_things.html#.Vxbz49R94rg

iot6.eu/sites/default/files/IoT6%20-%20D7.3.pdf

CHAPTER 3

The Technical and Business Innovators of the Industrial Internet

The advances in sensor technologies in recent times have been driven by the advent of high-speed and low-cost electronic circuits, a change in the way we approach signal processing, and corresponding advances in manufacturing technologies. The coming together of these new developments in these synergetic fields has allowed sensor designers and manufacturers to take a completely novel approach, such as introducing intelligence for self-monitoring and self-calibration, thereby increasing the performance of their technical products. Similarly, the advances in sensor manufacturing technologies facilitate the production of systems and components with a low cost-to-performance

© Alasdair Gilchrist 2016
A. Gilchrist, *Industry 4.0*, DOI 10.1007/978-1-4842-2047-4_3

ratio. This includes advances in microsystem technologies, where manufacturers are increasingly adopting techniques such as surface and bulk micromachining. Furthermore, initiatives exploring the potential in the field of digital signal processing involve novel approaches for the improvement of sensor properties. These improvements in sensor performance and quality mean that multi-sensor systems, which are the foundation of the Industrial Internet, can significantly contribute to the enhancement of the quality and availability of information. Due to these initiatives and an innovative approach by designers, this has led to new sensor structures, manufacturing technologies, and signal processing methods in individual and multi-sensor systems. However, it is the latest trends in sensor technology that have the most relevance in the Industrial Internet and these are the miniaturization of sensors and components, the widespread use of multi-sensor systems, and the increasing availability of radio wireless and autonomous sensors.

Previously, sensors that where embedded into devices or systems had to be hard-wired or rely on the host system for communication. Remote I/O devices could provide the communication interface and intelligence to interface applications with sensors with no communication circuitry, but again they typically had to be hard-wired, which limited their deployment options in outdoor locations. Bluetooth and ZigBee are lightweight technologies that have transformed sensor design by providing an embedded miniature radio for short distance communication. ZigBee, which has found numerous applications in the Industrial Internet due to its ability to build mesh networks that can span wide areas, is more prevalent in industrial applications than Bluetooth, which has its roots in the mobile phone industry. Developers heavily utilize Bluetooth in mobile accessories, applications, and short distance communication. Furthermore, advances in low-power WAN radio technologies and protocols has enabled these to be embedded into sensors and remote I/O devices, which has facilitated their deployment outdoors. This is true even at great distances away from the host operation and management applications.

Miniaturization

However, all these communication technologies would be impractical were it not for the rapid advancement in sensor miniaturization. Miniaturization has progressed to the stage that the manufacturers of sensors can reduce them to be the size of a grain of sand. This means that sensors can now be embedded anywhere and in anything, such as the clothes we wear, the packaging of the food we eat, and even our bodies.

Embedding intelligence into the sensor has also accelerated the path to miniaturization, as has integrating multi-functions into the design, such as temperature and humidity. For example, manufacturers that produce sensors that come fully calibrated, temperature compensated, and amplified, reduce the number of components needed on the PCB, which helps reduce size and weight as well as cost.

Some examples of the scale of miniaturization and how it has enabled use-cases of the Industrial Internet are in the medical and health care industry. One such device is the humble reed switch. A reed switch is a passive component that requires no power or additional components to work, which is, as we will see, one of its great advantages. A reed switch senses the presence of a magnetic field when a magnet is nearby and closes its connections. Once the magnet goes away, it opens the connections. The problem is that it is difficult to miniaturize passive components and still get them to work. Consequently, reed switches were typically a minimum of 25mm long, but after miniaturization, they have scaled down to around 3mm. That might not sound like a lot but it has had a dramatic impact on their use in industry. Reed switches are used in industrial, medical, and aerospace designs, among others.

Two of the most critical areas for miniaturization are the electronic/semiconductor equipment testing market and medical devices. Reed switches are essential in semiconductors as they are required to switch digital pulses billions of times a second, and reed switches do this perfectly. Additionally, they also have a place in medical implants and they outfit pill cameras, defibrillators, glucose monitoring devices, nerve stimulation devices, and have many more in-body applications. Reed sensors are perfect for these applications, as they use no power. Unlike semiconductor-based sensors which require batteries, reed sensors can sit in the body for many years without the need for removal.

Another advance in multi-sensor systems came about through the success of complementary technologies and through their proliferation. This was due to the popularity and acceptance of technology such as smartphones, systems-on-a-board, and even systems-on-a-chip (SoC). These devices come packed with multi-sensors and the software to drive them. For example, an Apple iPhone, the Raspberry Pi, and the Arduino with extension shields all provide the tools to create multi-sensor devices that can sense and influence their analogue environment through their interaction with the digital world. The availability of these development kits has accelerated the design process, by allowing the production of proof-of-concept (PoC) models. They have driven innovation in the way we deploy multi-sensor devices into industrial system automation and integrate M2M with cyber-physical systems to create Industrial Internet of Things environments.

Cyber Physical Systems (CPS)

The Industrial Internet has come about due to the rapid advancements in digital computers in all their formats and vast improvements in digital communications. These disciplines are considered separate domains of knowledge and expertise, with there being a tendency for specialization in one or the other. This results in inter-disciplinary knowledge being required to design and build products that require information processing and networking; for example, a device with embedded microprocessor and ZigBee, such as the Raspberry Pi or a smartphone. However, when we start to interact with the physical world, we have a physical domain to contend with and that requires special knowledge of that physical and mechanical domain such as that of a mechanical engineer. Therefore, it is necessary to identify early in the design process whether the product is to be an IT, network, or a physical system—or a system that has all three, physical, network, and digital processing features. If it has, then it is said to be a cyber-physical system. In some definitions, the networking and communications feature is deemed optional, although that raises the question as to how a CPS differs from an embedded system.

Information systems, which are embedded into physical devices, are called "embedded systems". These embedded systems are found in telecommunication, automation, and transport systems, among many others. Lately, a new term has surfaced, the *cyber-physical systems* (CPS). This distinguishes between microprocessor based embedded systems and more complex information processing systems that actually integrate with their environment. A precise definition of cyber-physical systems (CPS) is that they are integrations of computation, networking, and physical processes. Embedded computers and networks monitor and control the physical processes, with feedback loops where physical processes affect computations and vice versa.

Therefore, a cyber-physical system can be just about anything that has integrated computation, networking, and physical processes. A human operator is a cyber-physical system and so is a smart factory. For example, a human operator has physical and cyber components. In this example, the operator has a computational facility—their brain—and they communicate with other humans and the system through HMI (human machine interface) and interact through mechanical interfaces—their hands—to influence their environment.

Cyber-physical systems enable the virtual digital world of computers and software to merge through interaction—process management and feedback control—with the physical analogue world, thus leading to an Internet of Things, data, and services. One example of CPS is an intelligent manufacturing line, where the machine can perform many work processes by communicating with the components and sometimes even the products they are in the process of making.

An embedded system is a computational system embedded within a physical system; the emphasis is on the computational component. Therefore, we can think of all CPS as containing embedded systems, but the CPS's emphasis is on the communications and physical as well as the computational domains.

CPS have many uses, as they can use sensors and other embedded systems to monitor and collect data from physical processes. These processes could be anything such as monitoring the steering of a vehicle, energy consumption, or temperature/humidity control. The CPS systems, unlike embedded systems, are networked, which allows for the possibility of the data being available remotely, even globally. In short, cyber-physical systems make it possible for software applications to interact with events in the physical world. For example, to measure peaks in energy consumption in an electrical power grid—the physical process—with which the CPS interacts through its embedded computation and network functions.

Unlike traditional embedded systems, which are often standalone devices with perhaps a communication capability built in, CPSs are designed to be networked with other complementary devices and so have physical I/O ports. CPS is closely related to robotics, and a robot is a good example of a CPS, as it has clear physical components that can manipulate its environment. Robots are good at sensing objects, gripping and transporting objects, and positioning them where required. In factories, robots are used to do repetitive jobs that often require heavy lifting or the positioning of large awkward items in an assembly line. Robots have computation, network, and physical components to enable them to run software to do their tasks, such as to read sensors data, apply algorithms, and send control information to servomotors and actuators that control the robots arms, levers, and mechanisms. Robots also communicate with back-end servers in the operations and management domain and with safety devices on the assembly line. An example of this is that in some deployments, such as in stock handling in warehouses where robots retrieve or return stock from shelves and bins, robots work very quickly. They move at tremendous speeds, performing mechanical arm actions in a blur, and do not tire or need rest breaks, so they outperform humans in every regard. However, robots and humans do not always work together safely, and it is necessary then if a human comes into the working vicinity of a robot that the robot must slow down and perform its actions at a speed compatible with humans. Consequently, robots work better and much more efficiently in human-free environments.

Robotics is an obvious example of a CPS but presently they are adapted to work in many IIoT use-cases, such as working in hazardous environments such as in fire fighting or mining, or doing dangerous jobs such as bomb disposal or performing heavy-duty tasks such as lifting car assemblies on the production line. However, other uses for other types of CPS abound, such as in situations that require precision, such as in automated surgery, and in coordination, in the case of air-traffic control systems.

Real-world applications in the Industrial Internet for CPS are mainly in sensor-based applications where network-enabled CPS devices monitor their environment and pass this information back to an application on another networked node, where computation and analysis will be performed and feedback supplied if and when required. An example of this is collision detection and protection in cars, which also use lane-change awareness systems, also driven by CPS.

It is envisaged that advances in physical cyber-mechanics will greatly enhance CPS in the near future, with improvements in functionality, reliability, usability, safety, adaptability, and autonomy.

Wireless Technology

Wireless communication technology's adoption into the enterprise had a somewhat inauspicious start back in the early 2000s. Deemed to be slow and insecure, many IT security departments shunned its use and others went further and banned it from the enterprise. Industry was not so quick to write it off though, as Wi-Fi had tremendous operational potential in certain industrial use-cases, such as in hospital communications, and in warehouses, where vast areas could not be easily covered by hard-wired and cabled solutions.

Gradually, wireless technology evolved from the early systems, which could only offer limited bandwidth of 1-2Mbps and often a lot less than that over limited distances of 50 feet to high-performance Gbps systems. The evolution in the technology was a gradual step-process, which took most of the decade where incremental improvements in performance, matched with improvements in security. Security was a major issue as radio waves are open to eavesdropping since they broadcast over the air and anyone listening on the same frequency can make them out. Additionally, access points broadcast an SSID, which is a network identifier so that the wireless devices can identify and connect to its home network. For the sake of convenience, early Wi-Fi access points were open with no user credentials required for authorization and data went unencrypted over the air or was protected by a very weak security protocol called WEP (Wired Equivalent Protocol).

These failings were unacceptable to enterprise IT, so Wi-Fi found itself a niche in the home, SMB (small medium business), and in some industries where security concerns were not such an issue. Industrial uses for wireless technologies such as Wi-Fi, Bluetooth (which has similar early security setbacks), and later ZigBee were concerned with M2M data communications over short distances within secured premises, so the risk of data leakage through windows from overpowered or poorly positioned access points and antennae was not a problem. Similarly, in vast warehouses, Wi-Fi was a boon for M2M communication with remote-control vehicles and the low speeds, low throughput, and poor encryption were not an issue. Therefore, wireless technology gained an initial niche in the industrial workplace that was to grow over the decade, and as speeds and security improved beyond all previous expectation, wireless communication had become a driving force and an enabler of the Industrial Internet.

This transformation came about because of technological advances in wireless modulation, which enabled more bits or symbols to be carried over the same carrier frequency signal. For example, Wi-Fi went from 802.11b with a realistic bit-rate of 1-2Mbps (theoretical 11Mbps) to 802.11n with realistic bit-rate of 70-90Mbps (theoretical 300Mpbs) in less than a decade. Significantly, security also had rapid improvements with the flawed WEP replaced eventually with WPA2, a far more secure encryption and authentication protocol. The combination of these improvements was Wi-Fi's redemption in the IT enterprise and it has now gained full acceptance. In some cases, it is the preferred communications medium, as it provides flexible and seamless mobility around the workplace.

Further amendments to the standards in 2013 have produced staggering results, with 802.11ac and 802.11ad producing theoretical bit-rates of 800Mbps and 6Gbps, respectively, due in part to advanced signal modulation through ODFM and the MIMO technology (Multi-IN/Multi-OUT). They use multiple radios and antennae to achieve full-duplex multi-stream communications.

Additionally, amendments to the 802.11 protocol in 2015 produced the 802.11ah, which is designed for low-power use and longer range. It was envisaged to be a competitor to Bluetooth and ZigBee. One new feature of 802.11ah, which makes it differ from the traditional WLAN modes of operation, is that it has predetermined wake/doze period to conserve power. In addition, devices can be grouped with many other 802.11h devices to cooperate and share a signal, similar to a ZigBee mesh network. This enables neighbor area networks (NAN) of approximately 1KM, making it ideally suitable for the Industrial Internet of Things.

However, it has not just been in Wi-Fi that we have experienced huge advancements; other wireless communication technologies have also claimed niche areas, especially those around M2M and the Internet of Things. Some of these wireless technologies have come about as a result of the need for improvements over the existing alternatives to Bluetooth and ZigBee, which were originally focused on high-end mobile phones and home smart devices, respectively, where power and limited range were not constraining factors. In the Industrial Internet, there are many thousands of remote sensors that must be deployed in areas with no power and are some distance from the nearest access point, so they have to energy harvest or run from batteries. This makes low-power radio communication essential, as changing out batteries would be a logistical and costly nightmare.

The wireless technologies to address the specific needs of IoT devices are Thread, Digimesh, WirelessHart, 802.15.4, Low-Power WiFi, LoHoWAN, HaLow, Bluetooth low-power, ZigBee-IP NAN, DASH7, and many others.

However, it is not just these technologies, which have accelerated the innovation that are driving the Internet of Things. We cannot overlook the platform, which coincidentally in many ways was introduced in 2007 with the release of the first real smartphone, the IPhone.

The iPhone is a perfect mobile cyber-physical system with large processing power. It is packed with sensors and is capable of both wireless and cellular communications. It can run complex apps and interact with devices and its environment via a large touch screen, which is an excellent HMI (human machine interface). The importance of the introduction of the smartphones (Google's Android was soon to follow) was that to both consumer and industrial IOT there was now a perfect mobile CPS that was ubiquitous and highly acceptable. The problem previously was how humans were going to effectively control and manage IoT devices. The introduction of the smartphone, and later tablets solved, that problem, as there was now a mobile CPS solution to the HMI dilemma.

IP Mobility

It was around 2007 that wireless and smartphone technology transformed our perception of the world and our way of interacting with our environment. Prior to 2007 there was little interest in mobile Internet access via mobile devices even though high-end mobiles and Blackberry handsets had been capable of WAP (Web Access Protocol). Device constraints and limited wireless bandwidth (2G) made anything other than e-mail a chore. The 3G cellular/mobile networks had been around for some time, but uptake was slow. That was to change with the arrival of the smartphone and the explosive interest in social media, through Facebook and the like. Suddenly, there was market need for any time, any where Internet access. People could check and update their social media sites, chat, and even begin to browse the Internet as fast data throughput, combined with larger touch-screen devices made the browsing experience tolerable. Little did we know at the time the disruptive impact that the smartphone and later the tablet would have on the way we worked and lived our lives.

Prior to 2007 and the advent of the smartphone and mobile revolution, IT governed the workplace as far as technology and employee work devices were concerned under the banner of security and a common operating environment. However, with the proliferation of employee's own smartphones and tablets coming into the workplace, things were going to change. Employees were working on their personal devices, iPhones, and Android phones that had capabilities that at least matched the work devices that they loathed. This ultimately led to the revolution of employees demanding to use their own devices to do their work, as they were more comfortable with the devices, the applications, and it was in their possession 24/7 so they could work whenever and wherever they wanted. This was referred to as BYOD (bring your own device) and it went global as a workplace initiative. Similarly, riding on the success of BYOD it became acceptable to store work data on personal data storages, after all it was little use to employees to have 24/7 access to applications, and reports but not data so BYOC (bring your own cloud), although not nearly so well published, as employees stored work data on personal cloud storage such as Box and Amazon, became ubiquitous.

However, most importantly is what these initiatives had achieved. They transformed the way that corporate and enterprise executives viewed IT and employee work practices. The common consensus was these initiatives fostered a healthier work/lifestyle balance, created an environment conducive to innovation, and increased productivity.

Regardless of the merits of BYOD, what it did was introduce mobility into the workplace as an acceptable practice. What this meant was IT had to make available the data and service to the employees even if they were working outside the traditional company borders. Of course, IP mobility was abhorrent to traditional IT and security, but they lost the war because innovation and productivity ring louder in the C-suite called security.

However, at the time little did anyone know the transformative nature of IP mobility and how it would radically change the workplace landscape. With the advent of IP mobility, employees could work anywhere and at any time, always having access to data and company applications and systems through VPNs (virtual private networks). Of course, to IT and security, this was a massive burden, and it logically led to deploying or outsourcing application in the cloud via SaaS (software as a service).

Make no mistake, these were radical changes to the business mindset. After years of building security barriers and borders, security processes and procedures to protect their data, businesses were now allowing the free flow of information into the Internet. It proved, as we know now with hindsight, to be a brilliant decision and SaaS and cloud services are now considered the most cost effective ways to provide enterprise class software and to build SME data centers and development platforms.

IP mobility is now considered a necessity with everything from software to telephone systems being cloud-hosted and is available to users anywhere they have an Internet connection.

An example of IP mobility is that employees can access cloud services and SaaS anywhere, which makes working very flexible. Previously with on-premises server-based software, employees could only access the application if they were within the company security boundaries, for example using a private IP address, within a specific range or by VPN from a remote connection. However, both of these methods were restrictive and not conducive to flexible working. The first method meant physically being at the office, and the second meant IT having to configure a VPN connection, which they were loathed to do unless there was a justifiable reason.

Cloud-hosted software and services get around all those barriers by being available over the Internet from anywhere. Additionally, cloud-hosted services can integrate easily through APIs with other cloud-based applications so employees can build a suite of complementary applications that are tightly integrated, thus making their work experience more efficient and productive.

Network Functionality Virtualization (NFV)

Virtualization is a major enabler of the IoT; the decoupling of the underlying network topology is essential in building agile networks that can deliver the high performance requirements in an industrial environment. One of the ways to achieve this is through flexible network design where we can remove centralize network components and distribute them where they are required as software. This is the potential offered by NFV for the Industrial Internet, the simplification, cost reduction, and increase efficiency of the network without forsaking security.

NFV is concerned with the virtualization of network functionality, routers, firewalls, and load-balancers, for example, into software, which can then be deployed flexibly wherever it is required within the network. This makes networks agile and flexile, something that traditional networks lack but that is a requirement for the IIoT. By virtualizing functions such as firewall, content filters, WAN optimizers for example and then deploying them on commodity off-the-shelf (CotS) hardware, the network administrator can manage, replace, delete, troubleshoot, or configure the functions easier than they could when the functions were hard-coded into multi-service proprietary hardware.

Consequently, NFV proved a boon for industry, especially to Internet service providers, which could control the supply of services or deny services dependent on a service plan. For example, instead of WAN virtualization or firewall functions being integrated into the customers' premise equipment (CPE)—and freely available to those who know how to configure the CPE—a service provider could host all their virtual services on a vCPE.

Here lies the opportunity—NFV enables the service provider to enhance and chain their functions into service catalogues and then offer these new features at a premium.

Furthermore, NFV achieves this improvement in service provisioning and instantiation by ensuring rapid service deployment while reducing the configuration, management, and troubleshooting burden.

The promise of NFV is for the IIoT to:

- Realize new revenue streams
- Reduce capital expenditure

- Reduce operational expenditure
- Accelerate time to market
- Increase agility and flexibility

Increasing revenue and decreasing provisioning time, while reducing operational burden and hence expense are the direct results NFV.

NFV is extremely flexible in so much as it can work autonomously without the need of SDN or even a virtual environment. However to deliver on the promise, which is to introduce new revenue streams, reduce capital and operation expenses, reduce time to market for services, and provide agile and flexible software solutions running on commodity server hardware, it really does need to collaborate with and support a virtualized environment.

In order to achieve agile dynamic provisioning and rapid service deployment, a complementary virtualization technique is required and that is network virtualization.

Network Virtualization

Network virtualization provides NFV with the agility it requires to escape the confines of the network edge and the vCPE; it really is that important.

Network virtualization provides a bridged overlay, which sits on top of the traditional layer-2/3 network. This bridged overlay is a construction of tunnels that propagates across the network providing layer-2 bridges. These tunnels are secure segregated traffic flows per user or even per service per user. They are comparable to VLANs but are not restricted to a limit of 4,096, instances. Instead, they use an encapsulation method to tunnel layer-2 packet flows through the traditional layer-3 network using the WxLAN protocol.

The importance of this bridged overlay topology (tunnels) to NFV and the IIoT is that it provides not just a method for secure multi-tenancy via a segregated tunnel per user/service, but also provides real network flexibility.

What this means in practical terms is that it doesn't have to connect a physical firewall or load balancer inline on the wire at a certain aggregation point of the network. Instead, there is a far more elegant solution.

An administrator can spin up, copy over, and apply individual NVFs to the specific customer/service tunnel, just as if they were virtual machines. This means there is terrific flexibility with regard to deploying customer network functions and they can be applied anywhere in the customer's tunnel.

Consequently, the network functions no longer have to reside on the customer's premises device. Indeed some virtual network functions can be pulled back into the network to reside on a server within a service provider's network.

Network virtualization brings NFV inside the CSP network onto servers that are readily accessible and easy for the CSP to manage, troubleshoot and provision. Furthermore, as most of the network functionality and configurations are carried out inside the Service providers POP there are no longer so many truck-rolls to customers' sites. Centralizing the administration, configuration, provisioning, and troubleshooting within the service providers own network greatly reduces operation expense and improves service provisioning and deployment, which provides agile and flexible service delivery.

One last virtualization technology plays a part in a high performance network that cannot be ignored—that is the Software Defined Network (SDN).

SDN (Software Defined Networks)

There is much debate about the relationship between NFV and SDN, but the truth is that they are complementary technologies and they dovetail together perfectly. The purpose of SDN is to abstract the complexities of the control plane from the forwarding plane.

What that means is that it removes the logical decision making from network devices and simply uses the devices forwarding plane to transmit packets. The decision-making process transposes to a centralized SDN controller.

This SDN controller interacts with the virtualized routers via southbound APIs (open flow) and higher applications via northbound APIs. The controller makes intelligent judgments on each traffic flow passing through a controlled router and tells the forwarding path how to handle the forwarding of the packets in the optimal way. It can do this because, unlike the router, it has a global view of the entire network and can see the best path to any destination without network convergence.

However, another feature of SDN makes it a perfect fit with the IIoT and that is its ability to automate, via the SDN controller, the fast real-time provisioning of all the tunnels across the overlay, which is necessary for the layer-2 bridging to work.

SDN brings orchestration, which enables dynamic provisioning, automation, coordination, and management of physical and virtual elements in the network. Consequently, NFV and SDN working in conjunction can create an IIoT network virtual topology that can automate the provisioning of resources and services in minutes, rather than months.

What Is the Difference Between SDN and NFV?

The purpose of SDN and NFV is to control and simplify networks; however they go about it in different ways. SDN is concerned primarily with separating the control and the data planes in proprietary network equipment. The rationale behind decoupling the forwarding path from the control path is that it bypasses the router's own internal routing protocols running in its control plane's logic.

What this means is the router is no longer a slave to OSPF or EIGRP algorithms, which are the traditional routing mechanisms that determine the shortest path between one routing host to another in order to determine the most efficient or shortest path between communicating nodes. These algorithms were designed for a more peaceful and graceful age.

Instead, the SDN controller will assume control. It will receive the first packets in every new flow via the southbound OpenFlow API and determine the best path for the packets to take to reach the destination. It does this using its own global view of the network and its own custom algorithms.

The best path an SDN controller takes will not necessarily be based on the shortest path like most conventional routing protocols instead the programmer may take many constraints into consideration such as congestion, delay, bandwidth as it is designed to be programmable.

Smartphones

At the heart of all the recent trends in IoT and machine learning is the smartphone, and everyday we see innovations that center on the device as a controller, a system dashboard, and a security access key, or a combination of all three, that enable myriad of applications and analytic tools. The smartphone, because it has huge consumer adoption and market penetration (at levels in excess of 90% in developed countries), enables IoT innovation. Indeed, it is because people always have their smartphones in hand that mobile banking and NFC cardless payments have proliferated.

An example of the smartphones' importance to IIoT innovations is that it is the primary human machine interface (HMI). Consider Ford's innovative approach to in-car infotainment systems to see how industry is approaching future design. Car manufacturers design and build a car to last a first time owner 10 years; however designing and constructing the body shape and the underlying mechanics is troublesome enough without having to consider the infotainment system, which is likely to be out of date within five years. The solution that Ford and other car manufacturers came up with was to supply a base system, a visual display, and sound system, that integrates with a smartphone through a wireless or cable connection and via software APIs. By doing this, Ford circumvented the problem of the infotainment system being outdated before the car, after all the infotainment system now resides on the owner's smartphone and an upgrade is dependent on a phone upgrade. The point here is that it would only be through a commonly held item, one that the driver would likely always have in their possession, that this design would be feasible. It would not work with for example a laptop.

Similarly, though just a project just now, Ford is looking at drive-train control for their cars. What this would mean is that instead of Ford building the same class of cars but with economic, standard, or sports variants, they could produce one model, and the owner could control the drive-train via a smartphone application. Therefore, the family car could be either a sedate economic car for school runs or a high-performance gas-guzzler on the weekends, depending on the smartphone app. The outlook here is that cars would not become commodity items but their performance could be temporarily altered by a smartphone application to suit the driver or the circumstances.

Smartphones appear to be the HMI device of choice for IoT application designers, as can be seen in most remote control applications. The smartphone is certainly the underpinning technology in consumer IoT where control and management of smart devices is through a smartphone application rather than physical interaction. However, it is not as simple as convenience or pandering to the habits of the remote control generation. Smartphones are far more intelligent than the humble remote control and can provide much more information and feedback control.

Take, for example, the IoT capabilities of a smartphone. A modern Android or iPhone comes packed with sensors, including an accelerometer, linear acceleration sensor, magnetometer, barometer, gravity sensor, gyroscope, light sensor, orientation sensor, among others. All of these sensors, in addition to functional capabilities such as a camera, microphone, computer, storage and networking, can provide the data inputs to IoT applications and subsequent information regarding the phones environment can be acquired, stored, analyzed, and visualized using local streaming application tools.

Smartphones are not just HMI or remote controls—they are sensors, HMIs, and application servers and they can provide the intelligence and control functions of highly sophisticated systems, such as infotainment and drive-train technology. However, smartphones will only be at the edge of the proximity and operations and management domains, as they are not yet or likely to be in the near future capable of handling Big Data, pentagrams of unstructured and structured data for predictive analysis.

Deeper analysis, for example predictive analysis of vast quantities of Big Data, will still be performed in the cloud, but importantly fast analysis of data and feedback that is essential and required in real-time by industrial applications will be performed closer to the source and high-performance local servers are presently the most likely candidate.

However, the embedded cognitive computing ability in a smartphone will advance in the coming years, taking advantage of the sensors and the data they produce. Streaming analytic algorithms will enable fast fog-like analysis of sensor data streams at local memory speed without recourse to the cloud. As a result, smartphones will act as cognitive processors that will be able to analyze and interact with their environment due to embedded sensors, actuators, and smart algorithms.

An example of Industrial Internet is in retail. Smart devices with the appropriate apps loaded determine the location of a customer in a supermarket and spy on he or she is viewing. This is possible via RFID tags on products that have very short range, so their phone will only detect the products directly in front of the customer. The app will be able to describe through the display or importantly through the speaker—for those visually impaired—to the user what he or she is viewing. For example, the type of product, the price, discount, and any calorific or nutrient data normally declared on the label. Knowing a user's location, activity, and interests will enable location based services (LBS), such as instantaneously providing a coupon for a discount.

The Cloud and Fog

Cloud computing is similar to many technologies that have been around for decades. It really came to the fore, in the format that we now recognize, in the mid 2000s with the launch of Amazon Web Services (AWS). AWS was followed by RackSpace, Google's CE, and Microsoft Azure, among several others. Amazon's vision of the cloud was on hyper-provisioning; in so much as they built massive data centers with hyper-capacity in order to meet their web-scale requirements. Amazon then took the business initiative to rent spare capacity to other businesses, in the form of leasing compute, and storage resources on an as-used basis.

The cloud model has proved to be hugely successful. Microsoft and Google followed Amazon's lead, as did several others such as IBM, HP, and Oracle. In essence, cloud computing is still following Amazon's early pay-as-you-use formula, which makes cloud computing financially attractive to SMEs (small to medium enterprises), as the costs of running a data center and dedicated infrastructure both IT and networks can be crippling. Consequently, many cash-strapped businesses, for example start-ups, elected to move their development and application platforms to the cloud, as they only paid for the resources they used. When these start-ups became successful, and there were a few hugely successful companies, they remained on the cloud due to the same financial benefits—no vast capital and operational expenditure to build and run their own data centers—but also because the cloud offered much more.

In order to understand why the cloud is so attractive to business, look at the major cloud providers' business model. Amazon AWS, Microsoft Azure, and Google Cloud dominate the market, which is hardly surprising as they have the data centers and financial muscle to operate them. Amazon, the early starter launching in 2005, built on that early initiative to build their cloud services with increasing services and features year after year. Microsoft and Google came later with full launches around 2010 to 2012, although with limited services. They have not wasted time in catching up and both now boast vast revenue from their cloud operations.

To explain how the cloud and fog relates to the Industrial Internet, we need to look to the services cloud providers deliver. In general, cloud providers dynamically share their vast resources in compute, storage, and networks among their customers. A customer pays for the resources they use on and 10 minute or hourly basis, depending on the provider, and nothing else. Setup and configuration is automatic and resources are elastic. What this means is that is you request a level of compute and storage and then find that demand far exceeds this. The cloud will stretch to accommodate the demand without any customer interaction; the cloud will manage the demand dynamically by assigning more resources.

There are three categories of service—IaaS (Infrastructure as a Service), PaaS (Platform as a Service), and SaaS (Software as a Service). Each category defines a set of services available to the customer, and this is key to the cloud—everything is offered as a service. This is based on the earlier SOA (service orientated architecture), where web services were used to access application functions. Similarly, the cloud operators use web services to expose their features and products as services.

- IaaS (Infrastructure as a Service)—AWS's basic product back in 2005 and it offered their excess infrastructure for lease to companies. Instead of buying hardware and establishing a server room or data center a SME could rent compute, storage, and network from Amazon, the beauty being they would only pay for what they used.

- PaaS (Platform as a Service)—Came about as Microsoft and others realized that developers required not just infrastructure but access to software development languages, libraries, APIs, and microservices in order to build Windows-based applications. Google also supplies PaaS to support its many homegrown applications such as Android and Google Apps.

- SaaS (Software as a Service)—The precursor to the cloud in the form of web-based applications such as Salesforce.com, which launched in 1999. SaaS was a new way of accessing software, instead of accessing a local private server hosting a copy of the application, users used a web browser to access a web server-based shared application. SaaS was slow to gain acceptance until the mid 2000s, when broadband Internet access accelerated, thus permitting reliable application performance.

In context to the Industrial Internet, the cloud offers affordable and scalable infrastructure through IaaS. It also provides elasticity in so much as resources can scale on demand; therefore there is no need to over-provision infrastructure and networks, with the cloud you can burst well beyond average usage, as the cloud is elastic; it assigns resources as required, albeit at a price. Similarly, the cloud providers offer virtual and persistent storage, which is also scalable on demand. This is a major selling point for cloud versus data center deployments as the capacity planning requirements for the data storage of the industrial Internet can be vast.

For example, an airliner's jet engines generate terabytes of data per flight, which is stored onboard the aircraft and sent to the cloud once the aircraft lands, and that is just one aircraft.

Therefore, having elastic compute and storage facilities on demand but only paying for the resources used is hugely financially attractive to start-ups even large cash rich companies. Additionally, PaaS provides huge incentives for IIoT, in so much as the cloud providers can supply development environments and tools to accelerate application development and testing. For example, Microsoft Azure provides support for .NET applications and Google provides tools to support its own in-house applications such as Big Data tools and real-time stream processing.

From a network perspective the major cloud providers, Amazon, Microsoft and Google provide potentially millions of concurrent connections, and Google run their own fiber optic network, including their own under-sea cables.

The cloud is a huge enabler for the Industrial Internet as it provides the infrastructure and performance that industry requires but is at the same time financially compelling. However, there is one slight problem. Latency, which is the time it takes data to be transmitted from a device and then be processed in the cloud, is often unacceptable. In most cases, this is not an issue as data can be stream analyzed as it enters the cloud and then stored for more thorough Big Data analytics later. However there are some industrial use-cases where real time is required, for instance in manufacturing. In some, if not most, instances within manufacturing, a public cloud scenario would not be acceptable, so what are the alternatives?

- Private cloud—An internal or external infrastructure either self managed or managed by a third party but with single tenancy that is walled off from other customers.

- Public cloud—A community that shares all the resources based on a per-usage model; resources are supplied on-demand and metered. This is a multi-tenancy model with shared resources, such as storage and networking; however, tenant IDs prevent customers viewing or accessing another customer's data.

- Hybrid cloud—A combination of the private and public clouds, which is quite common due to security and fears over sensitive data. For example, a company might store its highly sensitive data in a private internal data center cloud and have other applications in AWS.

- Multi-cloud—For example, a company might have applications in AWS and developers working on Windows in Azure with Android developers using Google Cloud, and other IT applications stored on other public clouds.

It is likely in Industrial Internet context that a private cloud would be more attractive as it is inherently private, although it still leaves the dilemma of latency, jitter, and packet loss if the private cloud is on the Internet. The alternative of a private cloud hosted internally is also fraught with difficulty. To host a private cloud requires one of three methods—host the cloud on an existing infrastructure and manage it in-house, host the cloud on existing infrastructure and outsource the management to a third party, or outsource the cloud management to a third party on the Internet.

There is a fourth way, which is to use open source software. OpenStack can be downloaded and installed though it takes skill and patience. It is not recommended unless there is in-house cloud skills and a deep understanding of each business unit's application requirements. Remember, by setting up a private cloud on in-house infrastructure, the effect is to virtualize and share all resources. No longer will HR's application run happily in splendid isolation on their dedicated server, and the same goes for manufacturing's ERP server and customer cares, CRM, and VoIP. But what happens when you start sharing all the resources?

In addition, private cloud implementations will be costly, time consuming, and, unless diligently deployed, potentially insecure. So what is the alternative for industrial applications that require low latency and deterministic performance?

The Fog

Cloud systems are generally located in the Internet, which is a large network of unknown network devices of varying speeds, technologies, and topologies that is under no direct control. As a result, traffic can be routed over the network but with no quality of service measures applied, as QoS has to be defined at every hop of the journey. There is also the issue of security as data is traversing many autonomous system routers along the way, and the risk of confidentiality and integrity being compromised is increased the farther the destination is away from the data source.

IIoT data is very latency sensitive and requires mobility support in addition to location awareness. However, IIoT benefits from the cloud model, which handles data storage, compute, and network requirements dynamically in addition to providing cloud based Big Data analysis and real-time data streaming analytics. So how can we get the two requirements to coexist?

The answer is to use the *fog*.

The fog is a term first coined by Cisco to describe a cloud infrastructure that is located close to the network edge. The fog in effect extends the cloud through to the edge devices, and similar to the cloud it delivers services such as compute, storage, network, and application delivery. The fog differs from the cloud by being situated close to the edge of the proximity network border, typically connecting to a service provider's edge router. It will be on the service provider's edge router that the fog network will connect to, thereby reducing latency and improving QoS.

Fog deployments have several advantages over cloud deployments, such as low latency, very low jitter, client and server only one hop away, definable QoS and security, and supporting mobility location awareness and wireless access. In addition, the fog does not work in a centralized cloud location, but is distributed around the network edge, reducing latency and bandwidth requirements as data is not aggregated over a single cloud channel but distributed to many edge nodes. Similarly, the fog avoids slow response times and delays by distributing workloads across several edge node servers rather than a few centralized cloud servers.

Some examples of fog computing in an IIoT context are:

- The fog network is ideally suited to the IIoT connected vehicles use-case, as connected cars have a variety of wireless connection methods such as car-2-car, car-2-access point, which can use Wi-Fi, 3g/4G communications but require low latency response. Along with SDN, network concepts fog can address outstanding issues with vehicular networks such as long latency, irregular connections, and

- high packet loss by supplementing vehicle-vehicle communications with vehicle-infrastructure communication and ultimately unified control.
- Fog computing addresses many of the severe problems cloud computing has with network latency and congestion over the Internet; however, it cannot completely replace cloud computing which will always have a place due to its ability to store Big Data and perform analytics on massive quantities of data. As Big Data analytics is a major part of the IIoT and then the cloud, computing will also remain highly relevant to the overall architecture.

Big Data and Analytics

Big Data describes data that is just too large to be managed by traditional databases and processing tools. These large data structures can be and usually are made up of a combination of structured and non-structured data from a variety of sources such as text, forms, web blogs, comments, video, photographs, telemetry, GPS trails, IM chats, news feeds, and so on. The list is almost endless. The problem is with these diverse data structures is that they are very difficult to incorporate or analyze in a traditional structural database. Companies however need to analyze data from all sources to benefit from the IIoT, after all knowledge, such as customer trends and operational efficiency data can be distilled from all sorts of data.

However, in the IIoT the concern will be in handling vast quantities of unstructured data as well as M2M sensor data from thousands or more devices. Therefore, in order to gain value from this data there has to be an alternative way to handle and manage it.

Companies such as Walmart and Google have been processing Big Data for years and mining valuable hidden correlations from the data, but it has been done at great expense and with vast arrays of server and storage technology. However, they have undoubtedly been successful in their pursuit of handling and analyzing all the data they can retrieve from their operations. The Industrial Internet will require a similar approach as data from thousands of sensors will require managing and processing for valuable insights.

In industry, particularly in manufacturing, health services, power grids, and retail among others, handling and managing vast amounts of sensor data is nothing new, they have managed there production or services like this for years. For example in production, a sensor detects an event and sends the appropriate signal to an operational historian, which is a database that logs and stores data coming from sensors. The data stores are optimized to perform time-dependent analysis on the stored data by asking questions such as, how did this hour's production deviate from the norm? This database system manages this through complementary software tools designed to provide reporting and to detect trends and correlations.

Technology is capable of collecting sensor data from hundreds of sensor types and is developed to survive in hostile environments and to store data in the event that the database becomes unavailable. This is the long-established method for handling server data, so how will this change in the Industrial Internet?

The recent advances in sensor miniaturization and wireless radio technology have created a huge surge in the deployment of sensors, and consequently in sensor data. These advances led to the introduction of micro-electro-mechanical systems (MEMs). Sensors are now small enough to be deployed anywhere and can communicate over wireless technology. This has resulted in an explosion of data travelling from sensors to systems and sometimes back again, which is way beyond the levels of a few years ago. Now IIoT is seen as a major contributor of Big Data and as such requires the modern technologies to handle huge data sets of unstructured and dirty data.

Fortunately, for industry, cloud services are available to manage Big Data, with unlimited storage on-demand and open source technologies such as Hadoop, which is an open source cloud-based distributed data storage system optimized to handle unstructured and structured data. Similarly, there are tools for analytics such as MapReduce, developed by Google for its web search index. Hadoop utilizes its own file-system HDFS and works by assigning chunks of data to each server in its distributed storage system. Hadoop then performs a MapReduce operation before retrieving the results back into HDFS. This method is great for batch-job analytics; however, many IIoT use-cases will require fast real-time or close to real-time analytics on the data as it is in flight.

Therefore, knowing which technologies are needed depends on the type of Big Data, which can have several characteristics, termed the four Vs. They are each discussed next.

Volume

The ability to analyze large volumes of data is the whole purpose of Big Data. For example the larger the data pool, the more we can trust its forecasts. An analysis on a pool of 500 factors is more trustworthy than a pool of 10.

Velocity

Velocity is concerned with the speed the data comes into the system and how quickly it requires analyses. Some data, such as M2M sensors, will require in-flight or in-memory analysis; other data may be stored and later analyzed once in Hadoop. An example of a long-standing usage for high-velocity analysis is stock market and financial data. Financial institutions and banks have been analyzing this type of data at velocity, even going to the lengths of running a private submarine cable between exchanges in London and New York in order to shave a millisecond of the handling time of this valuable high-velocity Big Data.

Data velocity in the IIoT context, or streaming data as it is known, requires real-time or close to real-time as possible handling and analysis. This constraint puts additional pressures on the data storage and handling systems. The problem is the way the Industrial Internet tends to work; devices send sensor data back to an operations and management domain for processing. Now that data is typically sent to indicate a change of status with an entity or condition being monitored, and the sending device might be expecting a response.

This method of control feedback is very common in the industry and the system processing the data must be able to handle the data streams arriving from device sensors, process the data in flight (in memory), and identify and extract the data it requires, before it can take an appropriate action. For example, say a sensor on a high-speed motor within a centrifuge sends data that it has detected dangerous temperature, and simultaneously other sensors monitoring the motor report erratic performance and vibration. The system would want to know about this immediately, not as a result of a batch-job, but in real time, so that it could react and send a feedback signal to shut down the errant motor.

Variety

Another characteristic of Big Data is that it is typically messy and comes from a variety of sources, such as raw sensor feeds or web service APIs that do not fit neatly into organized relational structures, hence the need for NoSQL databases. As a typical use of Big Data processing is to extract meaning from unstructured data so that it can be input as structural data into an application and this requires cleaning it up. Sensor data is notoriously dirty as timestamps are often missing or lost in communications and therefore requires considerable tidying up before processing.

An example of this real-time insight into Big Data handling of sensor data is found in Smart City projects. For example, if a traffic monitoring system detects congestion or an accident from its roadside sensors, it can instantaneously send control feedback to change traffic lights, thereby easing traffic flows to reduce congestion.

Veracity

The problems with Big Data appear when we go beyond collecting and storing vast amounts of data and analyze the data stores using the 3 Vs and consider, is the data actually *true*.

The problem is that data is not only dirty or unreliable, it can be downright false. For example, say you harvest data from multiple sources of dumb sensors. You aggregate this data and transform it into information, on the basis that data leads to information that leads to knowledge. If the data was worthless to begin with, the results will be as well (garbage in, garbage out, as they say).

Value

Since not all data is equal, it becomes pertinent to decide what data to collect and analyze. For example, it has become a popular practice within industry and enterprises to collect everything, indeed the Big Data idea is to store everything and throw nothing away! The problem here is that data is valuable only if you can determine its relevance to the business value. After all having a Big Data value set means nothing unless the scientific data analysts have programmed software to retrieve the value from it. You must know what you are looking for. Big Data is not going to produce correlations and trends unless the algorithms have been programmed to search for such things.

Visibility

Visualizing data is hugely important as it allows people to understand trends and correlations better. Visualization software can present data in many formats, such as dashboards and spreadsheets or through graphical reports. Whatever way it is presented, it will still visualize the data in a human readable format making it easier to understand.

However, sometimes visibility means sharing data among partners and collaborators, and that is both a good thing and potentially hazardous. In the context of Industrial Internet it would be strange to offer information to potential competitors, as it could lead to others stealing highly sensitive data. For example, a Lathe machine in a factory will be programmed with a design template that determines the design and production of a specific product. To allow that information to leak to the Internet could be disastrous to the company. Say you contract an offshore company to produce one million shirts. Now that they have the template, what is stopping them from running off 2 million extras to sell on the black market?

The big point about Big Data is that it requires vast amounts of intellect to distil business value from it. Creating data lakes will not automatically facilitate business intelligence. If you do not know the correct question to ask of the data, how you can expect a sensible answer?

This is where we have to understand how or if machines think and collaborate.

M2M Learning and Artificial Intelligence

Big Data empowers M2M learning and artificial intelligence, the larger the pool of data the more trustworthy the forecasts—or so it would seem. M2M learning is very important and sometimes very simple, for example a multiple-choice exam. Say the exam wants to determine the student's knowledge level, so it might ask at random a question, it might be categorized as difficult, medium, or easy. If the student answers incorrectly the program might ask another question on the same subject at a different level. Its objective is not to fail the student but to discover the student's understanding of the topic. This is in simplistic terms called machine learning.

If we asked Google—with its vast arrays of computing power—to search for a nonsensical input, would we be expecting a sensible answer? The result is likely to be no. Moreover, here lies the problem, despite companies collecting, and harvesting vast quantities of data and constructing data lakes of unstructured data, how do they analyze this data in order to extract valuable information?

The answer is that currently they cannot, yet they can collect data in huge quantities and store it in distributed data storage facilities such as the cloud, and even take advantage of advanced analytical software to try to determine trends and correlation. However, we are not able to actually achieve this feat now, as we do not know the right questions to ask of the data. What we will require are data scientists, people skilled in understanding and trolling through vast quantities of unstructured data in search of sense and order, to distinguish patterns that ultimately will deliver value.

Data scientists can use their skills in data analysis to determine patterns in the data, which is the core of M2M communication and understanding, while at the same time ask the relevant questions that derive true value from the data that will empower business strategy.

After all a wealth of unstructured data—a data lake—means nothing unless you can formulate the correct questions to interrogate that vast data source in order to reveal those hidden correlations of potential information that add value to the company's strategic plan.

Consequently, data scientists have become the most sought after skilled professionals in the business. The Industrial Internet cannot exist without Big Data and intelligent analysis of the data delivered, and that requires skilled staff that understand data, algorithms, and business.

However, putting aside fear of robots and malicious intelligent machines, we can clearly see that even today, we have Big Data and analytics that do enable AI and that does deliver business value. For instance, IIoT systems listen to sensors that can interact with their environment and they can sense and react quicker that any human can. These sensors are our eyes, ears, nose, and fingers in the industrial world allowing us to proactively and reactively respond to our environment.

The huge benefits are that adding M2M communication with real-time analytics creates a cognitive computing system that is capable of detecting or predicting flaws, failures, or anomalies in the system that a human operator could not detect.

There are various types of machine learning or artificial intelligence which have ever-shifting definitions. For example, there are three general classifications of artificial intelligence, the classical AI approach, simple neuron networks, and biological network neuron networks. Each has its own defined characteristics.

Consider the classic AI approach that has been ongoing since the 1960s; you can identify the scientific approach is to identify the intelligence that humans find easy. For example, classic AI strived to find ways for machines to mimic human ability with regard to speech, facial, and text recognition. This approach has met with fairly mixed results, and speech and text has been far more successful than facial recognition, especially when compared to human performance. The problem with the classic AI approach is the machine's performance had to be judged and to be corrected by a human tutor so that it learned what was correct and what wasn't.

An alternative approach, again from the 60s and 70s, decided on a neural network, which was to mimic the way that the human brain works. In this scenario, the machine learns without any human intervention; it simply makes sense of the data using complex algorithms. The problem was that this required the machine to process vast quantities of data and look for patterns and that was not always readily available at the time. We have since discovered that simple neuron networks are a bit of a misnomer as it has little comparison to real neuron networks. In fact, it is now termed deep learning, and is suitable for analysis of large, static data sets.

The alternative is the biological neural network, which expands on the neural theme and takes it several steps further. With this AI model, the biological neural network does actually try to mimic the brain's way of learning, using what is termed spaced distributed representation. Biological neural networks also take into consideration that memory is a large part of intelligence and that it is primarily a sequence of patterns. Furthermore, learning is behavior based and must be continuous.

Initially, we might think well, so what, until that is we see how each model is used:

- Classic AI—This model is still used as it is very efficient in question answering, such as IBM's Watson and Apple's Siri.
- Neural networks—Data mining in large static data sets with the focus on classification and pattern recognition.
- Biological neural networks—It has many uses typically in security devices tasked with the detection of uncharacteristic behavior, as its strengths lies with prediction, anomaly detection, and classification.

How these would work in action is that each would take a different approach to solving a problem, let us say for example, inappropriate file access on a financial department's network. In this scenario, classic AI would report based on rules configured by an administrator, who shouldn't and who should have access and report any attempted violations. Now that works fine if it is black and white, some have access and others do not, but what if some people do need access but not every day?

This is where neural networks can play its part as it looks upon vast quantities of historical data and determines how often the network was accessed, how often, for how long and when, whether it was every day, week or just monthly.

A biological neural network takes it one step further; it doesn't just build a profile for the network, it builds a profile for each users behavior when accessing the network resource. It then can paint a picture of each user's behavior and determine anomalies in their behavior.

We must always consider artificial intelligence is not always the desirable goal and machine learning for the sake of it is not ultimately productive. A human is an astonishing machine, capable of learning and adapting to their sometimes hostile and complex work requirements and environments. For example, humans learn and can be taught skills far easier and cheaply than replacing hugely expensive robotic or CPS equipment on a production line. Humans also are fantastically capable of doing precise and delicate work, something robots are struggle with. Humans also have a brain, tremendous dexterity, strength, and commitment to workmates and family, unfortunately humans have a personality, are easily bored, and have an ego that no computer can possibly match. Therefore, repetitive, boring tasks are more suited to robots. After all humans, as brilliant machines as they are, were not designed to stand on a production line all day doing repetitive boring work. We have failings that make us human and these might be in the robot's favor in the job market of the future.

Augmented Reality

Augmented reality (AR), although still in its infancy, is stirring up quite an interest in the IIoT environment. Although AR is particularly new it was investigated as a futuristic technology decades ago and despite the obvious potential AR development fell by the wayside due to the lack of complementary technologies. However, in recent years interest in AR has been revived as all the complementary technologies are now a reality and in most cases thriving. Technologies such as AR visors, glasses, and headsets are now in production, and although still expensive from a consumer perspective, they are realistic for industry depending on the ROI (return on investment) of the use-case.

However, AR is not all about the glasses or visual display; it could just as well be a smartphone as it is also about data. After all AR is only as good as the information shadow that accompanies the object that you are looking at. An information shadow is the data that relates to the object that you are viewing, you could stare all you want at the walls of your house and you are not going to see the pipes and the wiring or anything different. For AR to work it needs the object you are studying to have a related 3D CAD diagram stored either locally if the AR device such as a tablet or AR headset can support large files or remotely in the cloud in the case of present models of AR glasses and visors. Then through the projection of the 3D CAD diagram, either through the heads-up display or onto the object itself, you will see pipes and wiring, and all that should be behind the wall.

Later versions of 3D do not solely rely on "as-built" CAD drawings as they can be notoriously poor, in the building and construction trades. Instead, they will rely on embedded sensors to transmit their location to build an interactive 3D drawing, which will show exactly what is located behind the wall. AR is extremely useful in the building and construction trade for planning work with the minimum of collateral damage.

AR has many other applications, one notable and passive form is interactive training of machinery maintenance. Previously technicians had to attend vendor training courses pass certification exams and develop skills over years of experience before they were capable of maintaining these machines or networks. Now with AR it is possible to accelerate the training because the in-depth data manuals and service guides are stored in the cloud along with 3D schematics and drawings. By projecting this information shadow alongside the AR view of physical machine, the technician can receive detailed troubleshooting steps to follow and the 3D image project onto the physical product will show them exactly where parts are located and how to access them. AR can play a massive role in industrial maintenance, removing most of the previous requirements for on-site expert knowledge, which due to its rarity was expensive.

Another use-case for AR is in control and management in industrial operation centers. In traditional operation centers, physical displays show displays of analogue processes, for example, temperature, pressure, RPM, among other sensor information. However, these physical dashboards were either physically connected or showed graphical representations of pre-configured and programmed dashboards. With AR, sensor data can be viewed from anywhere and projected onto any surface, thereby creating the facility to mix and match sensor data to be displayed in impromptu mobile dashboards.

Retail and the food and entertainment industries are also likely to be big players in the AR game, becoming early adopters and pioneers in building their own information shadow. For example, if the restaurant owners published to the cloud for public access, the prices, menu, and current available seating, then a potential customer when viewing the external facade will see that information superimposed through their AR glasses or device. Even if restaurant owners themselves do not take the initiative, social media may, and the potential customer will not see real-time data like the seating availability but they may see customer ratings and reviews.

The potential for AR and the IIoT is massive, so we have saved the most impressive use-case for last and that is AR use in emergency services. For instance, fire fighters already wear handsets and communication gear but, by using AR visors or helmets, they could be fed sensor information. This communicated environmental data, fed and displayed in real time to the fire fighter's heads-up display would provide vital information such as temperature, status of smoke detectors, and presence sensors. This information lets the firefighter have a heads-up view of the entire building floor by floor and know instantly if anyone is in the building and, if so, in what rooms, as well as the surrounding environmental conditions.

In short, only the boundaries of imagination and innovation of the developers and industry adopters of the technology limit the use-cases for AR when coupled with the Industrial Internet.

3D Printing

Additive printing or what is more commonly known as 3D printing is a major technology that enables the financial reality of the IIoT across many industrial use-cases. 3D printing works by creating an image as a computer file of either an existing product or through a CAD design one thin layer at a time. It builds on each subsequent layer until a full copy of the subject or CAD image has been completed. Once that computer file has been generated, it can then be fed to a 3D printer, which can interpret the coordinates to recreate the design using several techniques and substrates to create a physical representation of the subject.

3D printing enables a product to be created from a source file, which is not much different from how a programmable lathe machine creates a 2D product; however, it is the way that it does it in 3D that is different.

3D printing is therefore perfect for proof of concept, and modeling of theoretical designs as it is cheap and relatively quick. Most low-level industrial or consumer 3D printers work using plastic but higher-end printers can and do produce industry-quality components and products utilized in aircrafts, cars, and even in health care. These specialized components can be made from a variety of substrates, not just molten plastic, which is commonly used in consumer grade additive printing.

In industrial use, other techniques and materials are used to form and fuse the layers. For example, metal powder is used with binder jetting, which glues that metal powder together to form the 3D shape. In another industrial technique, glass, ceramic, and metal can be used as the base material using a technique, called power bed fusion, which used a high-power laser to fuse the material together to form the layers. There is also sheet lamination, which binds sheets of metal, paper, and polymers together, layer upon layer using force to create the binding. Metal sheets are then trimmed and cut by CNC milling machines to form the required shape.

The applications in industry are vast as 3D printing lends itself to all sorts of rapid prototyping, architecting, and construction. 3D printing enables not just rapid modeling but one lot sized production of customized objects as only the base software template file needs changed. This is very attractive in manufacturing where previously to change a products design required weeks of work refitting production lines and reconfiguring machines. With 3D printing, lot sizes of one can be entertained profitably and cost effectively.

It is important to differentiate between the media-hyped consumer market for 3D printing and the industrial reality. Components created using 3D printing in industry are not models or gimmicks as they are used by NASA, and in the aviation industry in jet engines. Similarly, they are commonly utilized in cars, with at least one manufacturer making their entire vehicle via 3D printing.

3D printing goes beyond just manipulating polymers, ceramics, paper, and metal—it also can be used in health care. Additive manufacturing is used in prosthetics and medical components such as medical sensors and actuators implanted within the body, such as heart pace-makers for example. However, the latest research is being driven by bio-medical requirements such as creating 3D printed skin, and other body tissue, and perhaps soon even complete organs. The goal here is to reproduce a patient's failing organs using 3D printing to create a replacement without the requirement of a donor transplant.

The way this works is that layers of living cells harvested from the patient are deposited via the 3D printer onto a culture plate to build up each layer of the three-dimensional organic structure. Because the 3D model is built from the patient's own living cells, the body won't reject the cells and the immune system won't attack them as a foreign entity, which is a huge problem with donor transplants.

People versus Automation

Humans are the greatest innovators and providers of technology and we adapt our environment to our liking so that we can exist in even the most hostile conditions. As a species, we can live in conditions as extreme as 50 degrees centigrade or survive for months in Antarctica in temperatures barely rising above -20 degrees. We can do this because, uniquely among animals, we can design and transform our environment to suit our physical requirements. Not only can we provide shelter for ourselves and kin in inhospitable climates but also we can produce food and sustenance through industry, agriculture, and hunting. As we have developed as a society, we have also developed social skills, rules, empathy, and emotional intelligence that allow us to operate as a social group living in harmony.

However, it is through competition that drives humans to excel. It is threat or rivalry that brings out our competitive nature, as we can see manifested in sport and war. We have egos and personalities, and we have emotions as powerful as love and as destructive as hate, envy, and greed. These may not necessarily be inhibitors to innovation or invention; they may actually be the spark that kindles the fire.

In times of strife or stress, humans are incredibly inventive, curious, and adventurous and many major innovations, inventions, discoveries, or social disruption has come around during periods of war or famine. Importantly, humans can realize not only their own physical limitations, which are great, but also can stretch beyond those boundaries and advance their capabilities and horizons even beyond their own planet.

We can develop robots and other cyber-physical systems to do our biding in environments too dangerous or ferocious for life. Furthermore, humans can adapt, learn, and change our behavior and physical characteristics to meet the requirements of our habitat.

The human body is an astonishing machine in its own right, as it can learn and counter physical stress, as we witness when undergoing any fitness-training regime. Our muscles adapt to the workload stress by getting stronger and developing stamina, as does our cardio-vascular system. Our heart rate lowers and our breathing becomes more efficient. However, most interestingly, our muscles and nerves create circuits that enable us to perform tasks without conscious thought. For example, a tennis player returning a serve with their backhand.

Machines and robots have none of these characteristics; they are simply mechanical devices designed to do a simple job, albeit repetitively and tirelessly. The fact that robots are indefatigable is of course their great strength, but they are extremely limited, or are at present. However, if robots or other cyber-physical systems could communicate through advanced M2M learning and communications, perhaps they could work collaboratively as a team, just like humans, but without the unpredictable performance driven by negatives of envy, bickering, bullying, moaning, or the more endearing positive qualities of camaraderie and team spirit displayed by human teams.

This is of course is the goal of machine learning and artificial intelligence—to create intelligent machines that can have some measure of cognizance. Ideally, they have a level of inherent intelligence that enables them to work and adapt to their and others circumstances and environment but without human attitude and personality flaws.

Currently in robotics, we are a long way from the objective; however, in software machine learning is coming along very well. Presently the state of machine learning and artificial intelligence is defined by the latest innovations.

In November 2015, Google launched its machine learning system called TensorFlow. Interest in deep learning continues to gain momentum, especially following Google's purchase of DeepMind Technologies, which has since been renamed Google DeepMind.

In February 2015, DeepMind scientists revealed how a computer had taught itself to play almost 50 video games, by figuring out what to do through deep neural networks and reinforcement learning.

Watson, developed by IBM, was the first commercially available cognitive computing offering. In 2015, it was being used to identify treatments for brain cancer. In August 2015, IBM announced it had offered $1 billion to acquire medical imaging company, Merge Healthcare, which in conjunction with Watson will provide the means for machine learning.

Astonishingly, Google's AlphaGo beat the world champion Lee Sodol at the board game GO, which is a hugely complex game with a best of five win. What was strange is that both Lee Sodol and the European champion (who had been beat previously by AlphaGo) could not understand Google's AlphaGo's logic. Seemingly, AlphaGo played a move no human could understand; indeed all the top players in the world believed that AlphaGo had made a huge mistake. Even its challenger the world champion Lee Sodol thought it was a mistake; indeed he was so shocked by AlphaGo's move he took a break to consider it, until it dawned on him the absolutely brilliance of the move. "It was not a human move … in fact I have never seen a human make this move". Needless to say, Google's AlphaGo went on to win the game. Why did AlphaGo beat the brilliant Lee Sodol? Is it simply because as a machine AlphaGo can play games against itself, and replay all known human games, building up such a memory of possible moves as a process of 24/7 learning that it keep can continuously keep improving its strategic game.

Google's team analyzed the victory and realized that AlphaGo did something very strange—it calculated a move, based on its millions of known human play training movements that a human player would only have had a one in ten thousand chance of recognizing and countering that seemingly crazy move.

In fairness to the great Lee Sodol, he did manage to outwit AlphaGo to win one of the best of five games, and that appears to be an amazing achievement.

References

http://www.wired.com/2016/03/googles-ai-viewed-move-no-human-understand/
www.ibm.com/smarterplanet/us/en/ibmwatson

http://www.wired.com/2016/03/googles-ai-viewed-move-no-human-understand/

http://3dprinting.com/what-is-3d-printing/ http://www.ptc.com/augmented-reality

https://www.sdxcentral.com/articles/contributed/nfv-and-sdn-whats-the-difference/2013/03/

CHAPTER 4

IIoT Reference Architecture

The Industrial Internet is reliant on the structure of M2M technology. Sometimes, it is older established technologies and practices that have been around for decades, that can spark innovation, and as a result, IIoT's architecture is often seen as a natural evolution of M2M. This is particularly true within manufacturing, which is the biggest user of IIoT technology, primarily due to its long history with machine automation, robotics, and M2M communication and cooperation.

Figure 4-1 shows a traditional M2M architecture and the IIoT architectural architecture. If you compare these, it becomes clear that the only difference at this high level is the addition of the Internet.

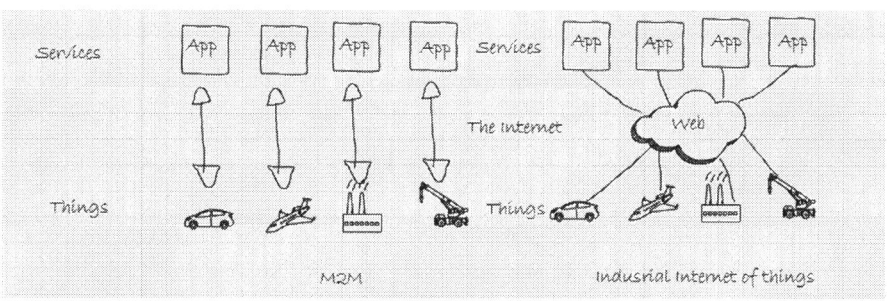

Figure 4-1. *M2M and IIoT architectures*

© Alasdair Gilchrist 2016
A. Gilchrist, *Industry 4.0*, DOI 10.1007/978-1-4842-2047-4_4

However, the reality when we dig deeper is quite different, as we will see when we consider the IIC (Industrial Internet Committee) reference architecture model.

The IIC Industrial Internet Reference Architecture

The Industrial Internet is, as we have seen, a subset of the broader horizontal classification called the Internet of Things. Whereas the Internet of Things encompasses everything—consumer, industry, enterprise, and commercial—the Industrial Internet takes a more focused view concentrating on energy, health care, manufacturing, public sector, transportation, and related industrial systems.

There are many industrial systems deployed today that are interconnected (M2M) and they combine a mixture of sensors, actuators, logic components, and networks to allow them to interconnect and to function. The difference with the Industrial Internet approach is that these industrial systems (ISs) will become Industrial Internet systems (IISs) as they become connected to the Internet and integrate with enterprise systems, for the purpose of enhanced business process flow and analysis. The IISs will provide operational data via its sensors to enterprise back-end systems for advanced data processing and cloud-based advanced historical and predicative analytics. The advanced cloud services will drive optimized decision-making and operational efficiencies and facilitate the collaboration between autonomous industrial control systems.

To realize these goals, IISs require a standard-based, open and widely applicable architectural framework. By providing reference architecture, the IIC (Industrial Internet Consortium) have provided the means to accelerate the widespread deployment of IISs, using a framework that can be implemented with interoperable and interchangeable building blocks.

The interchangeable nature of the Industrial Internet reference architecture is notable, as it is designed to be flexible and cover a wide range of deployment scenarios across many industries, such as energy, health care, and transportation, for example. Therefore, the reference architecture is a common framework that works at a high level of abstraction, which enables designs to follow and adhere with the reference architecture without burdening the design with unnecessary and arbitrary restrictions.

Furthermore, by de-coupling the architecture from the technical specifics and complexities the Industrial Internet reference architecture transcends today's available technologies. This approach will drive new technology development through the identification of technology gaps based on the architectural framework.

Industrial Internet Architecture Framework (IIAF)

Without going to deep into standards, it is necessary to briefly summarize some of the concepts of the IIAF as that will give you an understanding to the IIAF's structure. The IIAF is based on ISO/IEC/IEEE 42010:2011 standard, which codifies the conventions and common practices of architect design. By adopting the concepts of this standard, such as stakeholders and viewpoints, the IIAF follows the constructs of the specification, which is why the IIAF is described with reference to concerns, stakeholders, and related viewpoints.

A stakeholder is an individual, team, organization, or anyone who has an interest in the system. A concern refers to a topic of interest, and a viewpoint is a way of describing and addressing the concerns.

An example of a viewpoint and how it relates to a concern would be if you wanted to understand a system and how the machines interfaced with one another. The way the concern is addressed is through decomposition of the system design by drilling down to the specific interface details and characteristics of each inter-connected machine in the system. A component diagram would make the interfaces easier to understand and perhaps a sequence diagram would show how they interact. These diagrams address the concerns for that viewpoint (Figure 4-2).

Figure 4-2. Stakeholder PoV

Industrial Internet Viewpoints

Viewpoints come in four classifications—Business, Usage, Functional, and Implementation. The Business viewpoint addresses the concerns of stakeholders with regard to business-oriented matters, such as business vision, values, and objectives. Therefore, these viewpoints are of interest to business decision-makers. The Usage viewpoint addresses concerns related to how the system is expected to be used and is typically represented as use-case scenarios with users as actors interacting with the system to deliver a specific function. The Functional viewpoint focuses on the functional components in the system and how they inter-relate and integrate both internally and externally with other systems. Finally, the Implementation viewpoint it addresses the concerns of functional component lifecycles and how they are implemented within the system.

The Business Viewpoint

The Business viewpoint addresses business-oriented concerns such as how the system delivers value to the business and how it aligns with business strategy as well as financial concerns such as expected return on investment (ROI). Therefore, the stakeholders tend to be business visionaries and decision-makers who are able to identify, evaluate, and address business concerns and requirements. These stakeholders typically will have a major stake in the business and will have strong influence over decision-making and a deep understanding of company strategy.

The Usage Viewpoint

The Usage viewpoint takes the business requirements and realizes them through creation of user and system activities that deliver the required outcomes and business objectives. The activities described serve as input for the system requirements, which will ultimately enable the functionality required of the system.

The Functional Viewpoint

The functional Viewpoint addresses the stakeholders' concerns regarding the functionality of the Industrial Internet system. This is easier said than done in large, complex systems, so in order to address the concerns of IIS functionality, the IIC created the concept of the functional domain model. Furthermore, the IIC intended the concept of the functional domain to be applicable across a broad range of industries and their specific IISs requirements and specifications. Consequently, the application of the domain model is flexible and is not intended as either a minimum or a required set of functional domains. For example, the use-case in a certain industry may put more emphasis on some functional domains than others.

A typical IIS can be broken into five functional domains:
- Control domain
- Operations domain
- Information domain
- Application domain
- Business domain

These are shown in Figure 4-3.

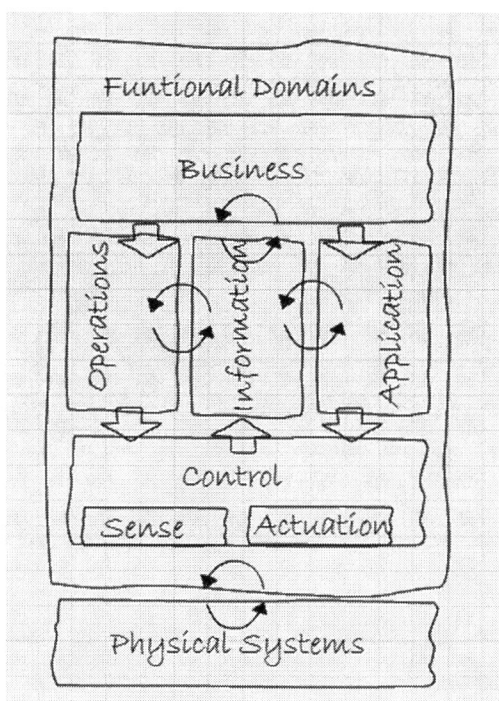

Figure 4-3. Functional domains

As the IIC designed functional domains to have broad applicability across an array of industrial use-cases, there are flexible implementation methods. For example, the design can combine functional domains into one single system, or split them up and distribute them across many systems.

Functional domains can be broken into sets of building blocks that are important to each functional domain, as experience has shown them to have wide applicability. However, these building blocks also are neither a complete nor a minimum set and a design can implement one or more into a system as required. The IIC's intention was to provide a starting point for conceptualizing the architecture of the IIS within the functional domains.

The Control Domain

A representation of the control domain is typically a collection of functional units that perform tasks such as reading data from sensors then logic units apply rules, logic, and subsequently applying feedback to the machines in order to exercise control over the process. In an industrial scenario, accuracy and resolution are both critical components of the control functions, and as such, the logic, the compute element, is usually situated as close to the sensors as is technically feasible. Examples of control domains may be in a large IIS system, for example, a control room in a nuclear plant or in smaller IISs a microprocessor in an autonomous vehicle, which controls temperature in a smart office.

The control domain is made up of a set of common functions, which may well vary in their complexity. An example could be that the IIS will require sensors and therefore the control domain will require a function to be able to read sensor activity. This could require not just hardware, software but also analytics, as there can be a requirement for recursive sensing, which is a complex feedback mechanism that requires real-time linkage to the IIS.

There is also a requirement for the control domain to intelligently manipulate actuation, which is the signals it sends to hardware, software via devices in order to control feedback to actuators, to trigger relays to initiate an action, for example, to return the system to safe operating limits. An example is a motor overheating, or sensors detect excessive vibration, which could prove to be dangerous.

Communication

The communication function connects all the functional sensors, actuators controllers, remote I/O devices, and ultimately the gateways to a common protocol. This is important, as in many IIoT scenarios such as in manufacturing, traditional field bus protocols are common, in which case translation is required between the sensors and the back-office applications. It is not always possible to implement one homogeneous system, for example a Greenfield site using IPv6 and CoAP. In many cases, legacy devices and protocols will be present such as ZigBee, in which case translation is required at the gateways. It is this entity abstraction, decoupling the existing devices protocol from the underlying complexity of the specifications and interface requirements, which makes inter-connectivity feasible.

For example, in a smart building there may be hundreds or thousands of devices from several vendors all using an array of protocols, but each highly specific to a task, that cannot be just cast aside. In order to incorporate them into the IIS, translational gateways will be required. Similarly, some dumb sensors, with no microprocessor or capability to support an IPv6 stack, will also require translation via an IPv6 proxy.

Modeling

The concept of data modeling, sometimes referred to as edge analytics, deals with the representation of the states, conditions, and behaviors of the system under control. The complexity required by data modeling depends on the system under control, for example determining the feedback requirement for a boiler to maintain an ideal temperature, or rather more complex to apply a behavioral template to an aircrafts jet engine. However, when we go above those requirements when dealing with systems dependent on artificial intelligence and cognitive capabilities, it becomes far more complex. Additionally, system complexity will require an understanding of peer systems and be capable of correlating data from both to have a holistic approach to controlling the system.

Asset Management

This is an enterprise operational function that enables control systems to easily onboard, configures, set policy, and performs self-diagnosis and automated firmware/software upgrades.

Executor

The Executor is one of the building blocks that's ultimately responsible for assuring policy; consequently, the executor allows understanding of condition and state. It also monitors the environment and the effects that may have on the system under control. The feedback control from the Executor can be sequential actions through its own actuation or via peer systems. There are two types of executor:

- *Straightforward*—A relatively simple control mechanism through feedback via an algorithm to control a boiler for example.

- *Sophisticated*—This technique incorporates aspects of cognitive and machine learning technologies that provide a high level of autonomy. That allows the machine to make its own best or worst decisions.

Chapter 4 | IIoT Reference Architecture

The Operational Domain

Traditionally control systems interconnected and communicated with only systems within their own environment, but with the advent of the Industrial Internet that has changed. Now IISs are communicating with other IISs on the other side of the globe, exchanging data and learning from each other. For example, optimizing a taxi company's operation may have obvious cost savings and lead to operational efficiencies.

The operation domain breaks down to the control domain as shown in Figure 4-4.

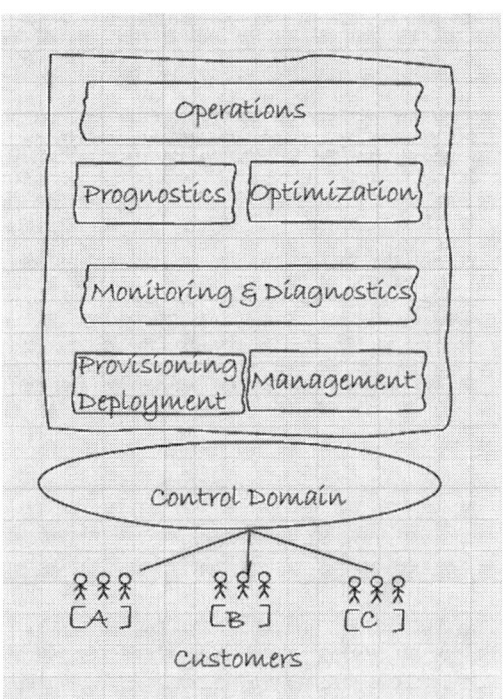

Figure 4-4. Operations to control domain

The operations domain consists of building blocks such as prognosis, optimization, monitoring and diagnosis, provisioning and deployment, and system management. These functions directly relate to the lower control functions and ultimately to the user. If we consider initially the task of provisioning and deployment, we can see it consists of a set of functions required to configure (provision a service) with regard to onboard, register, and asset tracking. The utility of the operational functions is to deliver assets online remotely, securely, regardless of scale and at to be cost effective.

The systems management utility offers a set of tools or functions that enable asset management functions to have bi-directional communications with assets (devices, actuators, and sensors) in order to control and manage remote assets.

Monitoring and Diagnosis utility is a valuable suite of functions that permit the detection and prediction of the occurrences of faults or problems. Having the ability to detect and proactively remediate problems is very important in industrial environments. Furthermore, monitoring and diagnosis have a dependency on Big Data historical and predictive analysis and real-time streaming forecasting.

Prognosis follows from the set of functions delivering monitoring and diagnosis as it serves as the predictive analysis engine on the ISSs. By consuming the vast data stored in Big Data lakes to determine historical and even predictive analytics, it is possible to identify potential issues with machinery and systems before they occur.

Optimization is a set of functions aimed at improving asset performance and efficiency. The ultimate goal is to improve asset reliability and performance while reducing energy consumption. Additionally, optimization strives to increase availability and efficiency as well as ensuring assets operates at peak optimal performance.

In order to obtain the goal of optimization, several conditions require to be reached:

- Automated data collection, validation, processing, and analytics.
- Capture and identify major system alarms, events, and issues such as power failures, system and network faults, latency, packet loss and downtime as well as non-responsive applications.
- The intelligence and capability to analyze and determine root causes for known problems.

The Information Domain

In the information domain, the objective is to transform the operation data harvested from the other domains, predominantly the control domain and its vast array of sensors, into information, which can be subsequently used as control feedback to stabilize or improve the process. However, in the information domain, the goal is not just to collect raw sensor data, the goal is to collect sensor data, store and process it as it can be transformed from raw data to information, and then distilled into knowledge. Transforming raw data into knowledge greatly enhances our ability to adapt and improve operational and business decision-making, and optimizing the process functions, which delivers greater productivity, profits, and operational efficiencies.

There are many examples of how information and knowledge of a process can deliver benefits, such as understanding the inverse relationship between operational costs and standard routines. If, for example, a fleet of trucks leaves the factory at 8:00 am each morning at the peak of rush hour, the fuel consumption and subsequent costs will reduce profits. By letting the trucks leave later, after the morning rush hour when congestion had subsided might be advantageous, but what about the return journey, when they again might hit the evening, and even more congested traffic? Information and analytics might solve this dilemma by suggesting an earlier departure, prior to the morning rush hour, which might relate to the driver being early for his delivery, meaning idle time. However, it would also facilitate an earlier return missing the evening rush.

Knowledge is always helpful as it allows managers to adapt to the present conditions, such as heavy snow that has closed normal routes, so they can reroute deliveries or be proactive in advising customers of imminent delays.

Data consists of functions for ingesting data and operational state from all the other domains; it is responsible for the quality-of-data processing such as data filtering, cleansing and removal of duplicate data. There are also functions that enable syntactical and semantic transformation, which relate to getting the format and the meaning of the message right, respectively. Another function of data is that it provides data storage, persistence, and distribution, and this relates to batch or streaming analytics.

All these functions can be performed in real-time streaming mode whereby the data is processed as they are received in quasi-real-time. However, this is not always possible or economical as commonly the data is stored for later batch job analysis. An example of the latter if where the vast data generated and collected from an airliners jet engines. This data is collected in-flight and due to the quantity and the cost of transmitting that amount of data over satellite links it is stored and uploaded to the system for later batch job analysis when the aircraft docks on landing.

Analytics contains a set of functions that enable data modeling, analytics and rule engines of advanced data processing. The analytic functions can be performed on/offline in streaming or batch mode. Typically, in streaming mode, the resulting events and alerts produced are fed into the functions of the application domain, while the results of batch job analytics go to the business domain.

The Application Domain

The application domain is responsible for the functions that control logic and deliver specific business functionalities. At this level, applications do not control machine or system processes, as that is the function of the control domain. Applications do not have the direct access, control, or the authority to control SSIs processes. Instead, application functions perform advisory functions; however, they do provide use-case rules and logic, as well as APIs whereby an application can expose its functionalities and capabilities to a system. The application domain as supports user interfaces that enable interaction with the application.

The Business Domain

The business domain enables integration and compatibility between Industrial Internet system functions and enterprise business systems such as ERP (enterprise resource management), CRM (customer relationship management), WSM (warehouse stock management), and many others. As an example, a predictive maintenance service for a fabrication yard may have historical and predictive analytics on past and probable failure rates on welding equipment, which allows them to forecast likely failures. By integrating their ERP systems, they can ensure stock is always at hand and reserved for the customer.

Implementation Viewpoint

The implementation viewpoint is probably the most interesting aspect of the IIRA as it involves technology rather than business or usage functions. Undoubtedly, IIS architecture will be guided by business and financial constraints, which will include, among other things, business requirements, alignment with corporate strategy, budget, and return on investment with regard to both financial and competitive advantage. However, from a high-level perspective it is interesting to see just how we go about building an IIS. What does it take and what are the inherent technical complexities? After all the hype over the IoT has been around for some years, so why are everyone not already doing it?

Well, interestingly, to answer that question, let us look at the implementation viewpoint.

Architectural Topology

The IIC has gone with traditional network architecture such as it follows a three-tier design, has gateway media-convertors at the border areas to convert and translate heterogeneous protocols and technologies.

Furthermore, it has a multi-tier storage topology that can support a pattern of performance tier, capacity tier, and archive tier that relates to the fog and cloud, respectively. In addition, the architecture also supports edge-to-cloud (fog) direct connectivity and a distributed analytics (cloud) topology.

The Three-Tier Topology

The best way to visualize an IIoT system is to take a brief look at the three-tier topology. This is a very simplistic representation of an IIS but it does represent the core areas that we need to be concerned with when designing an IIS network.

For example, the edge tier is the proximity network that hosts all those end-nodes, such as sensors, actuators, and remote I/O devices that provide communication facilities to otherwise dumb sensors and components. On first glance, this proximity network or edge tier looks simple enough; however, as we will soon see, the devil is in the details.

However, before we get ahead of ourselves let us run through the IIC's architecture, starting with the edge tier.

The Edge Tier

The edge tier is where data from all the endnodes is collected, aggregated, and transmitted over the proximity network to a border gateway. Depending on the protocols and technologies used within the edge tier, some data translation and interface integration may be applied at hubs, remote I/O devices, or protocol convertors. The edge tier contains the functions for the control domain.

The Platform Tier

The platform tier receives data from the edge tier over the access network and is responsible for data transformation and processing. The platform tier is also responsible for managing control data flowing in the other direction, for example, from the enterprise to the edge tiers. It is within the platform tier that we will locate the majority of the functions related to the information and operations domains.

The Enterprise Tier

The enterprise tier implements the application and business logic for decision support systems and end-user interfaces, such as for operations specialists. This tier hosts most of the application and business domain functions.

To get a clearer idea of how this all comes together, consider Figure 4-5.

Industry 4.0

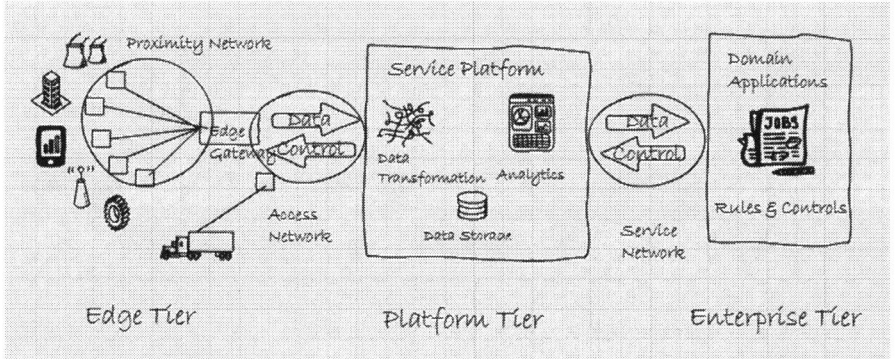

Figure 4-5. Three-tier architecture

In Figure 4-5, we can see the relationship between functional domains and the three-tier architecture. However, there is no exclusivity between a functional domain and a tier, as they can request services from one another. In addition, there are some functions, for example data transformation, which can take place in different tiers, such as the edge and the platform tiers. However, the data transformation in the edge tier is device or protocol specific, whereas in the platform, data transformation is a function performed on all the aggregated data.

Also shown are the scope and position of the proximity, access, and service networks. The three networks can take a variety of forms dependent on the geographical location an organizational topology.

Gateway-Mediated Edge

The purpose of the gateway-mediated edge device is to aggregate data flows and connection from all the endnodes and in a homogeneous network the gateway provides a single point of connection to a Wide Area Network (WAN). In a heterogeneous network, the gateway in addition to acting as an aggregation hub for the end devices may also have to provide media translation between differing standards such as RS232, Ethernet, or Modbus.

However, one of the benefits of using a gateway mediation device at the edge is that it provides for scalability by breaking down the complexity of the underlying endnode networks. By proving a single point of entry/exit into the edge tier and access to the endnodes beyond the gateway provides an ideal place for management and local data processing and analytics.

This concept of keeping some of the data processing an analytics close to the edge is termed *fog computing,* as opposed to cloud computing, as the data storage and processing is performed closer to the ground. Having the data analysis and processing performed close to the edge mitigates issues with latency as the data does not travel over long distances to a cloud provider. Instead it will be connected over an optimized link to a private or third-party network for processing.

Connectivity

How we connect endpoints, such as sensors and actuators, within the IIS can be a simple or extremely complex issue. If there is local LAN connectivity such as a field bus, Ethernet, or Wi-Fi, the issues are likely to be readily resolved. It is another matter however when connecting sensors in remote areas where power and communications options may be limited. Consequently, a vast array of communication technologies has arisen to try to fit one scenario or another, to the extent that the choice is bewildering. However, this ubiquitous connectivity is one of the enablers of the Industrial Internet so we should not complain.

The architectural role that connectivity plays is foundational in providing communications between endnodes and this plays a major role in interoperability, integration, and composability. Where integration requires common signaling and protocols, such as Ethernet, interoperability requires the same data formats (syntax) and a common method of understanding the meaning of the symbols in a message (semantics). Composability is the ability to understand what the intended message the user was trying to convey.

Consequently, when we speak of technical interoperability we mean the bits and bytes using a common interface (wire or wireless). Syntactic interoperability requires that a common protocol is used to communicate over the common interface.

IISs connectivity is comprised of two functional layers:

- Communication Transport Layer—Provides the means of communication between two endnodes as it provides for a common, interface, and protocol. Its role is to provide the technical compatibility and interoperability between two nodes. The functions relate to layers 1 – 4 of the OSI data model.

- Connectivity Framework Layer—The purpose of the functions in this layer is to provide the unambiguous format, syntax, and semantics of the common communication protocol. This function spans the upper layers 5 – 7 in the OSI model.

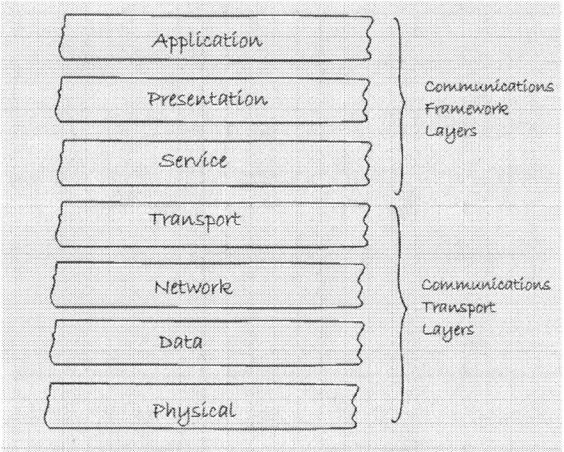

Figure 4-6. OSI communication layers

Key System Characteristics

Some of the key characteristics of the connectivity function are performance, determined by low latency (delay) and jitter (variation is delay), high throughput, resilience, and scalability. In IISs, latency and jitter are possibly the most important factors as IISs typically require very short reaction times as their actions are often tightly coordinated with real-world processes. Scalability is also a very desirable factor, as in some IIS use-cases millions of sensors endnodes may be connected so the IIS must be able to grow to accommodate them all.

Similarly, resilience is another major factor in an industrial context as reliability is paramount. The network must be robust enough to survive harsh real-world conditions. In this type of environment endpoint will fail, so localizing the loss and scope of communications in necessary. In addition, endpoints can restore themselves perhaps through energy harvesting so should be able to gracefully restore communications.

Connectivity Security

Data communications requires security that follows the three IT tenets of CIA (confidentiality, integrity, and availability). However, IISs security in the context of endpoints have the following requirements.

Explicit endpoint exchange policies via access management:

> Authorization techniques—This is through Identity and Access management (IAM) techniques and systems
>
> Encryption—Both for the identity authentication process and for the data inflight

In addition, as IISs can have long lifecycles up to 20 years in some cases the connectivity components should have longevity and be easily replaced or upgraded.

Key Functional Characteristics of Connectivity

The key role of the connectivity functions are to provide a common syntactic (speak the same protocol) and determine interoperability between endpoints so that they can understand each other. As a result, information must be structured so that each node can build its data in a common format and understand the exchange of information.

However, there is also delivery and permissions, which enable the intelligent decisions, discovery, and information exchange between nodes. As a result, a connectivity framework could support the authentication and authorization, for example, the right to read/write and exchange data.

However, in design topology, communications can be peer-to-peer, client server, or publish/subscribe. The publish/subscribe technology is a protocol based on a publisher that publishes date on a topic that a subscriber wished to enroll. The result is that it decouples the publisher from the subscriber, making them independent of one another. You will see many publish/subscribe models, such as MQTT and DDS, as you progress through the book.

There is also data quality of service functions, such as the requirement to deliver data at a higher priority, and with better quality of service. Data sent via an endpoint may be best effort data, e-mails, or data requiring high performance and low latency performance such as streaming video, data, or voice.

Therefore, we see the following delivery mechanisms:

> At most once delivery—This is commonly called fire and forget and rides on unreliable protocols such as UDP
>
> At least once delivery—This is reliable delivery such as TCP/IP where every message is delivered to the recipient
>
> Exactly once delivery—This technique is used in batch jobs as means of delivery that ensures late packets, delayed through excessive latency or delay or even jitter do not mess up the results

Additionally, there are also many other factors that need to be taken into consideration such as lifespan, which relates to the IISs to discard old data packets, much like the time-to-live factor on IP packets. There is also fault tolerance, which ensures that there is fault survivability and alternative routes or hardware redundancy is available, which will guarantee availability and reliability. Similarly, there is the case of security, which we will discuss in detail in a later chapter.

Key Functions of the Communication Layer

The communication layer functions can deliver the data to the correct address and application. If, for example, an application wished to communicate web browsing, it will address packets with the IP address of the recipient, such as 172.16.10.5, the question is how the recipient network card will know the application where those incoming packets should be delivered. The answer is quite simple, such as a postal service IP addressing sends data to an address, perhaps in the example 172.16.10.5. However, the port number, in this example 80 for HTTP traffic, tells the network interface card to which application the data must be delivered. This is similar to delivering mail to a condominium, where there may be hundreds of apartments all sharing the same address. The port (apartment number) ensures correct delivery.

However, there is a vast array of potential technologies and protocols. What is more, there are many ways that an endpoint can communicate via IP; for example endpoints could use unicast, which is point-to-point direct addressing (I know your individual address and port I will address the message specifically to you), or they can use multicast, where the source sends out a broadcast to a selected group of recipients. Furthermore, there is broadcast technique whereby a full broadcast to every node on the network segment receives the message whether they want the information or not. This is of course is hugely wasteful of bandwidth; however, it can be very efficient with regard to response times.

Some protocols, such as TCP/IP, are connection based in that they establish a three-way handshake to establish a session before they transmit data. TCP/IP was designed decades ago to facilitate data transmission over unreliable network media, such as those old 56Kbps dial-up modems. UDP on the other hand is unreliable, as it sends and forgets, but it is quick and extremely efficient over reliable networks such as LANs. There are also other modern protocols that have arisen to address the failings of TCP/IP and UDP and to address the modern technology requirements of specific technologies such as VoIP and streaming video. These new protocols use RTP (reliable transport protocol), which addresses most of the problems with long delays and setup time with TCP/IP, is it also mitigates the threat of the unreliability of UDP.

However, the problem is that most communication protocols are asynchronous, and this is a major issue. If we take for example a node communicating with a server sending alerts or alarms, it requires action in real or close to real time. Fortunately, field bus and other synchronous communication technologies deliver this, thereby delivering accurate communication. However Ethernet can be synchronous, either by design or by using timing protocols such as Precision Time Protocol or by deploying synchronous Ethernet interfaces.

In the last section, we went beyond design and physical connectivity, yet we must still consider that today's networks are also about efficient data management as data-handling techniques are required and relevant to how we reduce and transform data for input into analytics engines. Furthermore, managing vast quantities of data is what the IIoT is about and it is how we can efficiently communicate between vast numbers of devices.

Data Management

Handling the vast quantities of data that we expect to reap from Industrial Internet scenarios such as from in-flight data generated by an airliner's jet engine, or the motors of a wind turbine, requires special consideration. For example, we must consider how we can manage such vast quantities of data, an even if we can to what purpose. An oil drilling rig generates vast quantities of data, most of which was discarded as being worthless prior to the Big Data theory of keeping everything and throwing nothing away.

The problem, though, is that sensors produce vast amounts of data over time, and that data has to be handled, managed, and analyzed at great expense. Companies baulk at the expense of transporting and storing this data, as analysis is still dependent on data scientists asking the correct questions, yet they are doing so at increasing obscene levels. However, it is not stopping them hoarding data with little intention or strategy to derive value. Therefore, the industry must look for effective ways to reduce the sheer volumes of data they collect. Similarly, they should also look for efficient ways to analyze the data, thereby achieving optimal benefit from the data they do collect.

Data reduction tends to be through sampling or filtering, which reduces the volume and velocity of the data collected without losing the value of the information content. Sampling and filtering are data-reduction techniques devoid of any analytical factors. However, the actual analytical process summarizes data and produces approximations, so the data reduction is comparable to data compression, where you lose a lot of noise but the truth of the signal remains.

Importantly, if we are striving to harvest all the data of importance generated by an industrial process then we have to have an efficient way to identify and then collect the data we are interested in, and to ignore the remainder. The publish and subscribe pattern is an effective way to receive only information that we want to know about. Instead of receiving everything, we can selectively chose which topics or categories of data we wish to have forwarded to us. Publish and subscribe works extremely efficiently in OOP (Object Oriented Programming) and in managing the distribution of large volumes of diverse data across a network.

The technique of the publish/subscribe pattern is highly efficient for distributing and exchanging information updates between distributed systems that are loosely connected. The publish/subscribe model facilitates the optimization of communications across a network between those systems that produce content of interest, such as alarms and alerts, and those systems with an interest in hearing updates efficiently and as quickly as possible.

The advantages of the publish/subscribe pattern is that it can decouple the publishing and the subscribers in both location and time, which means subscribers immediately receive updates to a publisher's status regardless of their network location or time. This mitigates the risk of delayed alerts and alarms due to the subscriber polling for updates. However, publish/subscribe has other benefits, such as when they reside on an unreliable network, in which case the publisher will store and update the subscriber when connectivity is restored.

The purpose of publish and subscribe is to provide reliable low-latency data flow between edge tier nodes and cloud-based services. Conversely, it can also supply control flow data from systems in the platform and enterprise tier to endnode devices. Additionally, it provides scalability as it can handle vast amounts of publishers, which are data sources and subscribers, as data consumers.

Publish and subscribe can support the following modes of operation—alarm and event, command and control, and configuration. Depending on the protocol and the topology, it can work as a bus or broker topology. (We will discuss publish/subscribe protocols in depth later.)

Query

IISs have two ways to make software queries. There is the one-time query model, which is similar to the traditional database query, which is a batch-job process, and there is the continuous query. The later method is associated with data stream analytics and in-memory databases, which enable real-time data processing.

Queries can interrogate a select subnet of device-generated data. If they are directly addressable, the requested data is ether pulled by request or pushed to a gateway handling I/O for non-addressable devices. There is also selective use of the query command, which a user can manually query or automate through software. The latter method is typically utilized in batch-run analytics and reporting.

Storage, Persistence, and Retrieval

Storage, persistence, and retrieval serves a few important purposes, such as providing the functions that enable the creation of audit record, supporting simulations via replay, and providing reliable storage and scalability via cloud storage.

There are several forms of data storage supported within the IIS:

> Record—Supports defining a persistent subset of data in a sequential order, similar to a database. However, record data is not typically used as the subject of a query instead record data is used for record keeping and post-processing and analysis.
>
> Replay—This method supports retrieving data as a collection, which enables previously recorded data to be replayed as data-items in the order that they were received.
>
> Historian—Collects and stores data from local machine processes for batch processing.
>
> Big Data—The modern preferred solution for managing and storing voluminous quantities of system data. The ability to store vast quantities of data is down to the cloud's elasticity and cost-effectiveness.

Advanced Data Analytics

Like so many technologies or techniques in the Industrial Internet, advanced analytics is not something new; indeed the concept has been around for many decades and only the algorithms and methods have changed. Where the interest in advanced analytics has suddenly grown is with the introduction of Big Data and the Internet of Things. Previously, interest in advanced algorithms had been limited to certain business sectors such as insurance, marketing, and finance where risk and opportunity were the key drivers. However, with the advent of Big Data and the cloud resources to leverage it, has come interest from many other sectors in industry, in search of ways to improve decision-making, reduce risk, and optimize business outcomes.

This has brought about a seismic shift away from traditional business intelligence, which focuses more on descriptive analysis that uses historical data to determine what has happened. Instead, business leaders are looking to advanced analytics, which complements descriptive analysis by taking the analysis further via diagnostic analysis by asking, not just what happened, but why it happened. Furthermore, it uses predictive analysis to ask what will happen. And finally, it uses prescriptive analytics to ask what you should do (see Figure 4-7).

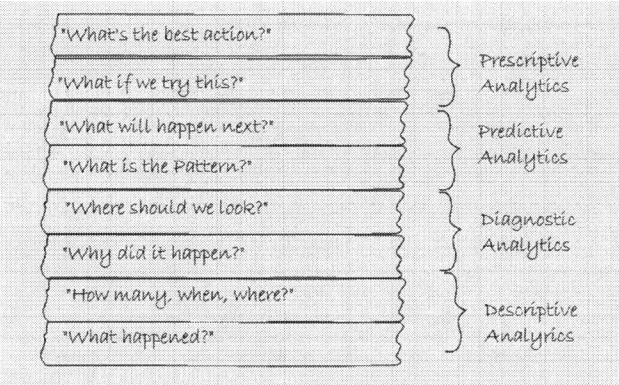

Figure 4-7. Types and methods of analytics

While basic descriptive analytics provide a general summary of historical data, advanced analytics delivers deeper data insight and ultimately knowledge from a more granular data analysis using a larger sample pool. The rewards of data-driven decision-making not only lowers risk and improves the probability of successful business outcomes, it can also uncover hidden correlations within the data. For example, oil rig applications apply analytics across a range of structured and unstructured sources—including geology reports, seismic studies of data, down-hole drilling data, satellite photographs, and sea-bed sonar data, among many other data sources—in order to recommend optimal exploration and drilling plans.

However, extracting value from data requires a range of skills and talents that cover data integration and preparation, creating database models and computing environments, data mining and developing specific intelligent algorithms. This is because making sense out of data is not a trivial task.

Volatile markets make it extremely difficult to predict future trends and the potential effect of business decisions. Having accurate data and reliable forecast allows business leaders to make confident decisions, such as when making financial planning and analysis (FP&A).

Standard business intelligence software tools can be used to analyze Big Data, such as predictive analytics, data mining, text analytics, and statistical analysis. Similarly, some enterprise BI software and data visualization tools can also prove valuable in the analysis process. However, one of the characteristics of Big Data is its mixture of structured and unstructured data and this semi-structured aggregated data may not fit well in traditional relational databases.

Furthermore, the computing resources available in traditional data warehouses may not be able to handle the processing demands of Big Data—for example, real-time data analytics of oil and gas pipeline data. As a result, a newer class of Big Data technologies such as Hadoop a distributed unstructured data storage solution and related tools such as YARN, MapReduce, and Spark. Similarly, NoSQL databases are a better fit than traditional SQL relational databases, as they can hold data in all structural formats. These new open source technologies provide the framework that supports the processing of large and diverse data sets.

References

The IIC Industrial Internet Reference Architecture

CHAPTER 5

Designing Industrial Internet Systems

Transducers (sensors and actuators) are the fundamental core of any M2M or IIoT architecture. They are the edge devices of any system and allow us to sense and manipulate our environment. However, transducers are only viable when they can communicate with other devices or a central processing machine and to do that they need to be interconnected.

As we saw earlier, M2M and the Industrial Internet of Things share a common reference and their architecture. Common to both are the area of the proximity network, which hosts the devices. The proximity area is key to the connectivity that interconnects all the devices and the backhaul to applications, back-end servers, and storage. Recall from the previous chapter the only real evolution between M2M and in an Industrial Internet is the sense that the IIoT has Internet connectivity.

Therefore, when we compare and study the architecture of M2M and the IIoT architectures, we can consider the Internet cloud system's intelligence and the interfacing middleware as part of the IIoT rather than an integral part of each IIoT or M2M system. Therefore, we can evaluate each system's back-end connectivity separately from the common proximity area.

© Alasdair Gilchrist 2016
A. Gilchrist, *Industry 4.0*, DOI 10.1007/978-1-4842-2047-4_5

Subsequently, if we consider the essential elements that make up the IIoT and the M2M architectures, which are fundamental requirements to build interconnected systems, and we can see these as being equivalent. This chapter covers the "Things"—the connection of devices that are the senses and manipulators of both the M2M and the IIoT system.

Devices can take many shape and forms, and this is a key prerequisite to understanding how to build interconnected systems. Not all devices are the same or indeed have the same complexity or connectivity requirement. Devices can be deployed in many ways—in clothing, credit cards, machinery, smartphones, jet engines, and even in your body—which means it's nonsense to try to create a one-fit system to monitor, manage, and control them.

In addition, there are an array of transducers to choose from, whether they are simple, cheap, passive analogue sensors, TTL digital sensors and actuators, complex embedded systems, advanced systems on a chip (SoC), or advanced custom proprietary logic built into industrial machines that support their own protocols. We cannot possibly have a single method of interfacing these. Furthermore, each device will have their own communication characteristics dependent on their role and this will determine how often and how much data they transmit. The role and type of device will have limiting physical characteristics such as size and power; for instance a System on a Chip (SoC) in a smart factory most likely has an external power source. On the other hand, a device deployed in a remote field location may be self-powered, perhaps by a small button cell battery, which will last not months but years. Of course, there are other alternative power sources, such as solar energy or wind power, but those still provide only very limiting power, and are unlikely to be suitable for powering a satellite or a 3/4G/LTE radio transmitter.

It's essential when designing interconnecting devices and systems that we take in consideration the physical characteristics and performance requirements of each device. These issues will determine which communication methods and protocols are suitable in each individual case. Designing interconnecting systems is not complex, and the correct tools for each case are already available. It is simply a case of matching the requirements to the available tools.

The Concept of the IIoT

There are at least four basic scenarios to consider when designing systems for connecting IIoT devices. Each use-case has a specific genre, and they arise from a single horizontal concept of the Internet of Things. The concept of the Internet-connected world of Things, which is the generic term for the IoT, is a wide-ranging horizontal technology that covers a multitude of disciplines.

Industry 4.0

Subsequently, more specific subsets of the concept arise as vertical markets and these are:

- Enterprise IoT
- Consumer IoT
- Commercial IoT
- Industrial IoT

Each of the scenarios has different characteristics based on the environment in which they are typically deployed and their own communication requirements. For example, in enterprise scenarios, IoT devices are typically connected using Ethernet (wired or wireless) connections, as it is the de-facto communication protocol in enterprises. Additionally, the devices will be only locally connected and communicate between gateways, back-end systems, and storage devices. When Internet connectivity is required by an enterprise IoT device, the entity will communicate via an Internet gateway.

Commercial IoT, on the other hand, such as power or water meters deployed in suburban or rural locations, will not have the benefit of locally available wireless or wired Ethernet and may require you communicate directly with a service provider's Internet gateway. This is similar to industrial IoT, but what defines IIoT, such as a smart factory deployment, is that many of the M2M communications are required, indeed mandatory, to be in real time. Therefore, to understand how to build and interconnect Industrial Internet device networks, you need a deep understanding of not just how the transducers will communicate with the rest of the world, and of the network's latency and predictability.

The Proximity Network

The choice of communication technology will directly affect the cost, reliability, and performance of the device-interconnecting system. However, as system designers can deploy devices in so many ways, no one technology is suited to all scenarios—there is no one size fits all.

As an example, consider a simple scenario of an IIoT deployment in a smart building. In this design, it would be a requirement to locate sensors and actuators throughout the building. There could be thousands of sensors required to monitor the environmental conditions, such as temperature, humidity, smoke, sound, or pressure, and perhaps hundreds of actuators to control the underlying systems. Physically wiring all these devices for power and communications in awkward places would be a very expensive, onerous, and time-consuming task, so deploying a battery and wireless sensor network for power and communications is an ideal solution.

Consequently, when we design the network, we must take into consideration the requirements of simple nodes that require low power yet need to communicate by transferring data over short distances. This is the basis of the WSN (wireless node network) local area network.

A WSN network consists of many WSN nodes, and these are usually cheap transducers that are networked to form a local mesh or star configuration that covers a local area such as a smart factory floor or the physical space in a smart building.

Each WSN node is like a cheap embedded sensor that has limited if any intelligence and performs a single function such as those used to monitor temperature or water levels. Being cheap, they can be deployed in large volumes within a Wireless Sensor Network (WSN), which covers a large strategic area. As a result, the WSN consists of a collection of distributed WSN nodes, such as sensors and actuators that are strategically situated and interconnected in a star or mesh topology. Due to the topology of the design, each WSN node relays data to its neighbor until the data reaches the edge node, which is typically a hub controller device. Return traffic, which performs a control function, will travel from the back-end application to the actuators in a similar fashion.

Another characteristic of WSN nodes is that they are low power so they can run off a battery or can harvest energy through solar, kinetic, wind, or electromagnetic radiation. WSN nodes generate or consume only small quantities of data; for example, they detect and communicate any changes in temperature or switch on a light or activate a relay. WSN nodes often use very basic communication interfaces and so they need to connect to an IP-aware edge device to communicate with the rest of the system.

WSN Edge Node

A WSN edge node is a gateway between the WSN network and typically an IP backhaul network. An edge node will also likely perform the translation and reframing of the local communication protocol between the WSN nodes and the edge node to IP, and it will perform aggregation and correlation of data from all the nodes. In addition, the edge node will be able to identify each WSN node through its standard interfaces. Additionally, the edge node will perform some local processing of the data and filter out much of the unnecessary traffic from traversing the backhaul IP network links. The reason we have this local processing at the edge device is that, in a WSN network, we only really have to know about a change in status of a sensor. There is no need to send a constant stream of identical data from a sensor all the way through the IP network to a control application if nothing has changed. Therefore, the edge controller only passes up notification of a change in status within a given sensor (delta). This greatly reduces the traffic passed to the controller without any hindrance to performance or efficiency.

WSN Network Protocols

A characteristic of all WSN nodes is that they require low-power communication technology and protocols. Wi-Fi is an obvious contender in some enterprise and commercial use-cases as it is ubiquitous in most commercial and industrial premises. However, Wi-Fi requires considerable power and may conflict with existing WLANs. Consequently, the IIoT devices used in industrial use-case are selected based on their radio's low-power consumption, and those that do not conflict with existing radio broadcasts. There are many competing technologies available.

Low-Power Technologies

The latest research and development of IIoT networking technologies is aimed at facilitating the development of low-cost, low-power connectivity solutions. These new low-power technologies support the design and creation of very large networks of intelligent and miniature WSD nodes. Currently, major R&D efforts include the research into low-power and efficient radios, which will allow for several years' of battery life. There is also considerable interest in the production of IoT devices capable of energy harvesting solar, wind, or electromagnetic fields as a power source, as that can be a major technology advance in deploying remote M2M style mesh networking in rural areas. For example, in a smart agriculture scenario. Energy harvesting IoT devices would provide the means through mesh M2M networks for highly fault tolerant, unattended long-term solutions that require only minimal human intervention

However, research and technology is not just focused on the technology. They are also keenly studying methods that would make application protocols and data formats far more efficient. For instance, low-power sources require that devices running on minimal power levels or are harvesting energy, again at subsistence levels, must communicate their data in a highly efficient and timely manner and this has serious implications for protocol design. There are also some serious trade-offs that need to be taken into consideration, for example higher data rates require higher power levels for the radios to transmit at higher frequencies, so lower frequencies are often considered to be better. However, lower frequencies also mean slower transmission, for the same amount of data, so the radio will stay on for longer, thereby consuming additional power. Therefore, we have to consider carefully the frequency (bandwidth), the size of the message being sent, and the technologies and protocols available in order to keep the radio power/transmit time to a minimum in order to preserve battery life.

Designing Low-Power Device Networks

The problem is with the IIoT deployments of transducers, not all technologies and protocols fit every scenario. There is also another problem in that there are technical issues that create confusion. As an example of this, consider a device (transducer) being connected via a mesh or star topology. How will this device communicate effectively with other devices or with the gateway edge device?

There are several problems here—one is the physical layer technology that is adopted, for example wired or wireless. Wireless technology—radio communications—must meet a required standard for the deployment. After all, it is of little use producing a startup product that uses a non-standard interface in the targeted industry. This is where standards are so important.

Now standards are hugely important, but very dry, so we will only cover the main industry standards that any industry specific product should comply with.

An example of this is that a startup developing a great product must adhere to existing relevant industrial standards or it will never be commercially accepted. Industry decision-makers are very risk averse and are unlikely to adopt a solution that does not meet existing industry standards.

As an example, consider a vehicle bus that has a specialized internal communications network that interconnects components inside a vehicle (car, bus, train, ship, or aircraft). Vehicles use special bus topologies, as they have specialist requirements for the control of the vehicle and for passenger safety, such as vehicle systems that require high assurance of message delivery to every component. Vehicle systems also require there to be non-conflicting messages, a deterministic time of message delivery, which necessitates redundant routing. However, low cost and resilience to EMF noise are two of the other characteristics that mandate the use of less common networking protocols.

Popular protocols include ControllerAreaNetwork (CAN), LocalInterconnect Network (LIN), and others. Conventional computer networking technologies (such as Ethernet and TCP/IP) are rarely used, except in aircraft, where implementations of the ARINC 664 such as the Avionics Full-Duplex Switched Ethernet are used. Aircraft that use AFDX include the B787, the A400M, and the A380.

Another well-established standard is IEEE 1451. The IEEE 1451, a family of Smart Transducer Interface Standards, describes a set of open, common, network-independent communication interfaces for connecting transducers (sensors or actuators) to microprocessors, instrumentation systems, and control/field networks.

This of course raises the question of what technology and protocols should be used, and to answer that question we need to understand—only at a high level—how devices (M2M) communicate.

To understand this, we need to go back to some basics—the OSI table for layered network communications (see Figure 5-1).

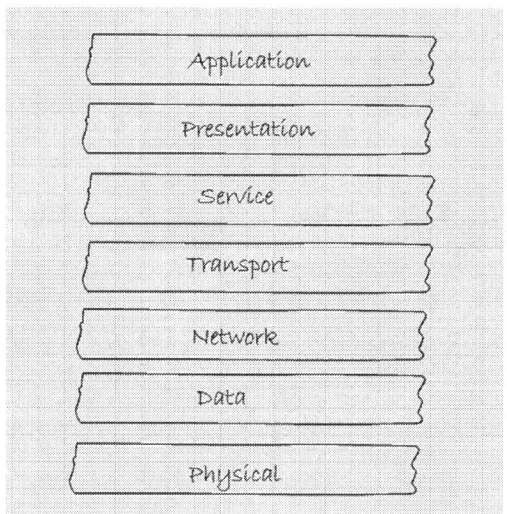

Figure 5-1. OSI table

The seven-layer OSI table, shown in Figure 5-1, determines each structural layer in data communications

With reference to Figure 5-1, we can isolate seven functional layers:

> Layer 1—The physical layer where hardware generates the electrical or optical signals to represent data traversing the physical network whether that be optical fiber, copper (cat 5 or 6), or wireless.
>
> Layer 2—The Data layer, the protocol that will enable each host system to understand the framing of the data. For example, what data pertains to host addressing, and what data is payload.
>
> Layer 3—The Network layer relates to the network address, which is IP specific, which identifies the host.

Chapter 5 | Designing Industrial Internet Systems

Layer 4—The Transport facility, which is TCP/IP specific and handles lost packets, out of sequence packets, and can request the sender to resend or reorganize packets based on their sequence number.

Layer 5—The Session layer maintains the session information between two hosts, which is especially important with TCP/IP as it can enforce encryption and other security measures.

Layer 6—The Presentation layer delivers the data in the format that the recipient requires.

Layer 7—The Application layer contains the end-to-end protocols used for communication. For example, the HTTP when a web client talks to a web server.

This was the ISO structure that served us well through the last few decades with data communications; however, web-based applications required a more condensed structure. That became the five-layer TCP/IP layered model, as shown in Figure 5-2.

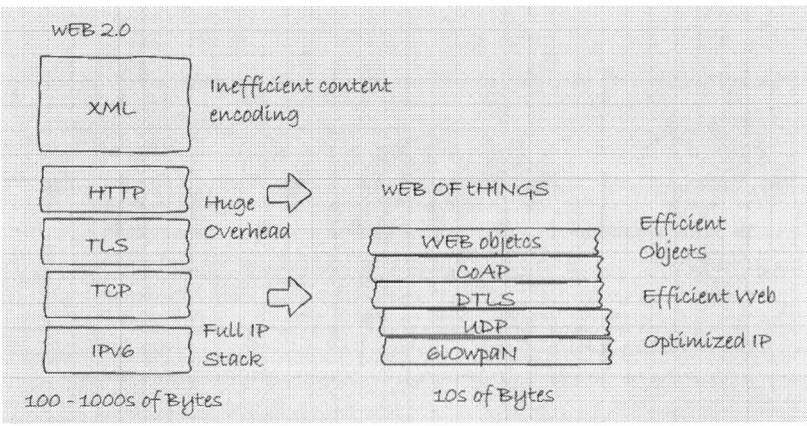

Figure 5-2. Web 2.0 layers

Similarly, the IIoT to achieve the performance required for real-time applications and services has required an even more condensed framework.

The point here is that data communications between endpoints has become more critical as we have adopted advanced technology and service level agreements. This is certainly the case with IoT, as real-time communication is vital to industry.

With the consumer IoT, which may or may not take off, it was not imperative to have real-time communication. Early adopters of the smart home, sport, or health wearable devices might be fed up with IoT automation failures but it was hardly catastrophic. However, in manufacturing or other industries, operational or system failure can cost millions of dollars in lost revenue. This of course makes the IIoT simultaneously hugely attractive but also a huge risk.

The question then is how has the IIoT addressed the problems, which may alleviate the stress of risk-adverse CEOs. The answer is through a different protocol structure, where as web-based applications reduced the ISO stack to five levels, the IoT structure has been reduced even more and it is dependent on vastly more efficient protocols.

For example, compare the models in Figure 5-3.

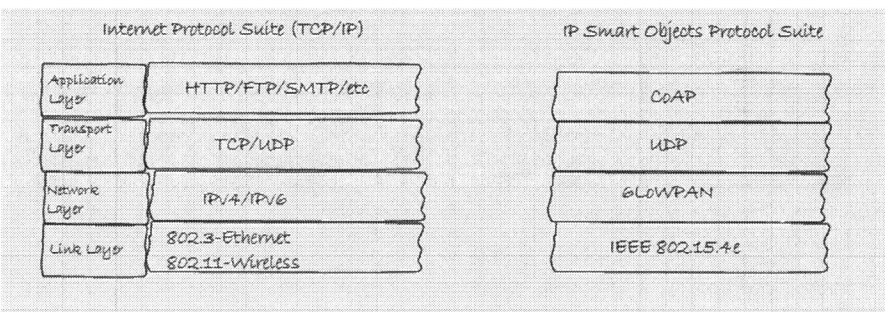

Figure 5-3. IP layers vs IIoT layers

The evolution of the OSI model clearly shows the major advances in technology and protocol efficiency. The IIoT embraces these challenges and utilizes the advanced physical technological and protocol communication advances to address existing IP issues with latency, jitter, and delay, which are the bane of traditional IP networks.

As a result, we now have a condensed OSI stack that provides more efficient node-two-node communication and delivers better performance, which can meet industry standards of efficiency. The result has been a major growth in radio and wired technology performance at the physical layers; after all, who could have believed a decade ago that we'd have consumer-based 1GB wireless and operators deploying 40/100 GB wired Ethernet technology in their backbone? Similarly, we have seen vast improvements in the efficiency of the data framing protocols—the ratio of overhead to data within an IP packet transmitted over low-power networks has plummeted.

Furthermore, we could not possibly have imagined the advances in technology and network protocols that would rise up to make the IoT feasible. The next section shows the vast array of technologies and protocols that are available when designing IIoT systems; just remember that one size does not fit all.

When we consider the Industrial Internet, it is not quite that simple to select one common protocol, such as IPv6, for the whole network for we must take into consideration how we will integrate the many diverse legacy communication technologies. This is important because many existing devices, sensors in particular, will be dumb or passive devices without the capability to host an IP stack, for example. These devices will probably have been configured to be directly connected via cables to a PCB or a PC perhaps using a RS232 serial connection. Many existing sensors in manufacturing environments will be configured this way and so we need to examine some of the existing communication technologies and how we can leverage them into the new Industrial Internet architecture.

Legacy Industrial Protocols

If we consider that many industrial technologies are implemented with up to 20-year lifecycles, then it is not strange to see that we still have many legacy technologies still in place. Indeed we have what would be considered ancient communication protocols still active in industrial environments; however, these protocols perform well and deliver the performance required in an industrial environment.

RS232 Serial Communications

Serial analogue communication and the RS232 specification go back to the days of mainframes and communications over modems that used standard telephone cables and connectors, nine-pin serial DIN plugs or even modems that housed a telephone handset. Generally, RS232 ran at 9,600Kbps but speeds have increased to 56Kbps and even higher. The RS232 specification is peer-to-peer where only two machines, the endpoint hosts, could communicate directly. This was simply because RS232 ran over a single cable such as a telephone line that connected the two hosts. There was also the fact that RS232 and the other serial communication modes ran at layer 1 of the OSI model. What this meant was there were no shared data link layer or MAC addressing, so hosts simply connected to one another once the baud rate, parity, and stop bits were determined to be compatible. Authentication was up to the host systems, not the communication technology. RS232 is therefore very simple and easy to implement on even the most physically constrained devices, as there is no need for processors and higher protocol stacks such as TCP/IP.

Devices running RS232 can easily be accommodated, despite their lack of IP capabilities via cheap serial-to-USB converters, which we can connect directly to the proximity area gateway. Additionally, being an analogue form of communication, RS232 can span much longer links than digital communication methods and is far less susceptible to noise and signal fade, even over long distances. Therefore, there is no compelling reason, if these devices are still performing the tasks they were designed for, to upgrade or replace them.

40-20ma Current Loop

Another traditional analogue method of transmitting data in industrial environments is through a technology called current loop, although technicians often refer to it as 4-20ma. The term 4-20ma comes from the way the technology works. A receiver on the loop will receive an analogue input from, say, a temperature sensor. The transmitter then modulates that analogue signal using a DC data source greater than 4ma and less than 20ma. This 16ma spread is split into tiny digital allocations and the current loop transmitter will assign the incoming analogue signal an appropriate level, proportionate to zero being 4ma and the highest possible value 20ma. By modulating the incoming signal on the DC current supplied by a power source—not the transmitter—an accurate analogue representation of the signal can be reproduced.

Current loop has all the benefits of other forms of analogue communication, for instance, it is not susceptible to noise and it can be transported long distances. Furthermore, because current loop uses a fake zero (4ma), any ground level noise is actually sitting below the signal, which has a base level of minimum 4ma. Therefore, any low-level noise is easily separated from the signal. Another great point about current loop is it is far easier to troubleshoot, as it follows electric laws and properties of a loop. At every point in a loop the current levels will be the same, which means it can easily be checked using just a digital voltmeter. If it shows 12ma at the source then it is guaranteed to show 12ma at the receiver.

Current loop is used extensively in legacy manufacturing environments and factories for connecting analogue devices. It is reliable, robust, and impervious to noise and interference.

Field Bus Technologies

The previous two examples of industrial technologies cited two slow but reliable analogue technologies that serve a purpose even in today's digital domains. However, these technologies are for handling sensor data in a non-critical scenario. Unfortunately, most manufacturing and processing systems in today's modern factories require real-time, or very close to real-time, performance. The way that this has developed over the years is to apply the concept of a distributed bus, where every node on the bus can communicate directly over the bus to every other node if permission is granted. This is not just how machines on factory floors communicate; it is also the foundation for automation, control, and management systems in cars, airplanes, trucks, and trains through the CAN (Controller Area Network), which was developed specifically to allow microprocessors in vehicles to communicate without the need of a local host computer.

However, CAN is not the only field bus technology. There are many other versions and standards such as EtherCat, Profibus, Foundation FieldBus, Modbus, BitBus, and many others.

Unfortunately, despite this wide choice of field bus technologies, interoperability is often not an option. This is because field bus technologies use layers 1, 2, and 7 of the OSI model. Layers 1 and 2 describe the physical characteristics of the technology such as bit timing, synchronization, how messages are formatted and assembled, and error handling, among other technical details. However, layer 7 deals with how to interact with the end application. Therefore, the goal to interoperability is to find a common application layer set of services that can be presented regardless of the lower technical specifications.

Despite this failure of interoperability, field bus implementations and standards grew popular for a number of reasons, cost being just one. Field bus typically is much cheaper to cable than 4-20ma installations, due to most devices sharing one common bus with individual drops. In addition to cost benefits is efficiency; a field bus distributed network can handle several variables per device, whereas 4-20ma can only handle and modulate one instance of a variable. Maintenance is another boon, because with field bus deployments it became possible for the first time to predict issues using the diagnostic data available of field bus devices. This provided a foundation for the predictive and proactive maintenance strategies that are often considered one of the drivers of an IIoT deployment.

Field bus is still around and will be for some time to come due to the long life expectancy of manufacturing equipment and the desire to preserve industry standards. However, field bus is being edged out by the advent of Industrial Ethernet.

Modern Communication Protocols

Industrial Ethernet

Industrial Ethernet is not that much different from the Ethernet we use in business. Industrial Ethernet uses the same frame and protocol, but the difference is it uses far more robust cabling and connectors. This effectively means that Industrial Ethernet is 100% interoperable with standard Ethernet at the data and protocol levels.

Industrial Ethernet uses robust cabling, as it must typically endure harsh environments with higher temperatures, greater levels of vibration, interference, and humidity. These harsh environmental conditions made standard Ethernet's category 5/6 cabling and cheap plastic RJ45 connectors unsuitable. The use of fiber optics could mitigate some of those problems such as interference.

Some early deployments of Ethernet ran into problems due to the collision detect and half-duplex nature of early specifications. However, full-duplex operation removes those early issues and Ethernet is now used in real-time deployments due to its inherent capability to communicate at the MAC address layer, making it quicker than TCP/IP.

The advantage of using a ubiquitous standard such as the 802.xx Ethernet standard protocol is that integration with IP networks such as the business LANs, WANs, and the Internet is trivial. Similarly, integration with business applications and services available in the cloud is also straightforward, as no gateways are required to transform or translate packets. They all can communicate over Ethernet. There is also the ability to connect to more devices at greater bandwidth and speed using larger data packages. However, the latter can also be seen as a drawback, as the minimum Ethernet packet is 64 bytes, whereas generally industrial systems tend to send data in small packets around 8-10 bytes.

Unfortunately, as we have seen, many of the Industrial Ethernet formats are non-interoperable. This is despite them all supporting the core Ethernet and TCP/IP layers from 1-4. It is again at layer 7, the application layer, and with the general communication models of encapsulation and management that the various implementations of Industrial Ethernet differ. For example, some are encapsulation models, such as Ethernet/IP, foundation field bus, and distribution systems such as Profinet.

With the encapsulation techniques, the existing field bus telegram packets are simply encapsulated into TCP or UDP containers. Once encapsulated, the TCP or UDP packets containing the field bus datagrams are transported across an Ethernet network as Ethernet frames, just like any other IP packet. The advantage of this method is that none of the underlying field bus technology has to be changed. All that is required is an appropriate field bus to Ethernet gateway. This method facilitates fast and efficient deployment without any engineering changes to the machinery or the PLC systems. In this configuration, Ethernet acts as a backbone, providing the heavy bandwidth and speeds that field bus technologies lack.

Handling time-critical traffic is a major concern in the Industrial Internet as most manufacturing machinery and processes are time-critical. Typical remote I/O Ethernet reaction times of around 100ms are unacceptable in manufacturing where deterministic times in the region of microseconds are required. Consequently, Industrial Ethernet must somehow address this issue and it can be resolved via several methods.

Encapsulated Field Bus or Application Protocol

One method to facilitate faster Ethernet reaction times for time-critical traffic is to encapsulate field bus traffic within TCP/UDP containers and tunnel that through an Ethernet network. This will not reduce the reaction time significantly, although it will bring it down to around 20ms on a local small segment. However, using synchronous Ethernet ports or using the precision time protocol (IEEE 1588), along with direct MAC addressing—using native Ethernet addressing techniques without any IP addressing—will bring the reaction time down to 1ms.

Another technique is to provide three channels. This works if you're sharing traffic over a segment, for example application protocols and data, non-critical time sensitive traffic, and critical-time sensitive traffic. Use standard TCP/IP for application traffic and a best effort delivery basis (5-10ms). Then, for the non-critical but still time-sensitive traffic (1ms), you can use prioritization at the data packet level to give these packets priority. Finally, for the time-critical traffic, we can install fast hardware switching at the Ethernet layer to bring times down to 0.2ms.

As you can see from the many formats available for Industrial Ethernet, there will be no one size that fits all and there will be cases where we will have to integrate field bus, 4-20ma, and RS232 into Ethernet in order to provide connectivity to the devices within the proximity network.

However, what we have discussed here is only the legacy and current communication technologies in manufacturing, processing, vehicle automation, and the grid, among others. We still have to consider the non-time-critical communication technologies that make the Internet of Things feasible. The trouble is there are even more of these wireless technologies, and no, there is no one solution that will fit all scenarios.

Standard Ethernet

The 02.3 Ethernet protocol, which is the network technology and protocol of choice for commercial and enterprise, is everywhere today. Ethernet is a protocol as well as a technology. It is transported over copper wire (cat 5/6 cables), optical fibers, and radio waves (Wi-Fi and microwave) in short, middle, and long-range deployments. Ethernet is popular because back in the early 1980s, it became the dominant networking technology in the growing commercial computer networking business. Ethernet was cheaper to cable than rivals such as TokenRing were, and easier for network designers to plan network topologies and for technicians to deploy and troubleshoot.

Consequently, Ethernet became the de facto networking technology in the enterprise and today you are unlikely to come across anything else in commercial or business computer networks.

The Ethernet protocol works on the idea that each network adapter installed in a device will have a unique MAC address, which is hard-coded or burnt into the device, making it unique. Because of this principle of uniqueness, each device can then be simply addressed using this MAC address. As a result, Ethernet encapsulates the data it wishes to send with frame headers, which are the source and destination MAC addresses of the sender and recipient machines. Ethernet can then pass this frame to the physical layer drivers to transport across the chosen media, copper, optical, or wireless. The recipient, like all other Ethernet devices on the network, listens for any traffic bearing its unique address and processes only the packets bearing its unique address, a broadcast address (a message for all devices), or a multicast address (a group address) if it is a subscribed member.

This makes Ethernet very straightforward to understand at this high level.

Ethernet does have some limitations, or at least it did in the early days, which made it unsuitable for industrial use. For example, early standards were half-duplex, which meant it could only transmit or receive one at a time, not do both simultaneously. In addition, it was based on shared media, so the network adaptor had to sense when there was no other communication and the media was available. As a result of this shared broadcast media and potentially large collision of domains, Ethernet required collision detection, which meant the network adaptor had to listen for and detect collisions and be prepared to retransmit the frames. These early designs were fine for enterprise and commercial use in small networks, which were the norm at the time, but were not suitable for industrial purposes that required more deterministic real-time performance.

Therefore, Ethernet was slow to gain acceptance in industrial applications, hence all these field bus technologies we have today. However, with Ethernet's stellar growth came vast technology improvements in the standards. Today, Ethernet no longer suffers from these earlier design limitations and full-duplex, advance Ethernet switching makes collisions redundant. Ethernet is also capable of working at baud rates up to 100Gbps over long-distance optic cables.

Consequently, Ethernet is finding its way out of the enterprise and into industrial use. It is already in use in DWDM long-distance core data transport networks, and it has usurped TDM technologies in radio backhaul for telecommunications in mobile operator networks (3G/4G/LTE).

Therefore, because of Ethernet's flexibility, ubiquity, and close relationship with the TCP/IP protocol suite, we can expect it to play a big part in the Industrial Internet.

Wireless Communication Technologies

The plethora of wireless communications over the last two decades has been one of the drivers for adoption of the Internet of Things by consumers and industries. Prior to wireless technologies becoming ubiquitous in the home, consumers had everything they required to build their own smart home; in fact, the technology has been around since the early 1980s. The problem was that there was no way to connect the sleek microprocessor controller to the sensors and actuators without running wires throughout the house. This of course would be very disruptive and potentially damaging, so was not worth the trouble. Wi-Fi, Bluetooth, and ZigBee have removed that large barrier to entry and now we can connect devices. However—and this is a recurring theme—there is not one wireless standard that fits all purposes. The array of wireless technologies has arisen because they serve different purposes. Some, such as Wi-Fi (802.11), are designed for high bandwidth and throughput for web browsing and downloading from the Internet. Bluetooth, on the other hand, was designed to provide a personal area network to connect wirelessly the devices that you wear or carry, such as a mobile phone or a watch to a headset, and it therefore requires short range and less power and throughput capability.

ZigBee's niche was connecting devices around the home so its range and capabilities fell somewhere between Wi-Fi and Bluetooth, or at least that is how it started out. Nowadays, ZigBee covers neighbor area networks (NAN) in industrial IoT scenarios and Bluetooth's range and capabilities have also improved beyond the scope of the PAN (personal area network). Regardless of the technology you choose to adopt, it will always have constraints, such as range, throughput, power, physical size, and cost. Therefore, you have to take a diligent approach when evaluating the wireless technology best suited for your needs.

IEEE 802.15.4

One of the early IoT enablers is the low-power IEEE 802.15.4 standard. It was first back in 2003 to set the standards for low-power commercial radios. The standard was upgraded in 2006 and 2011 to provide for even more energy efficient radio technology. IEEE 802.15.4 is the basis for the ZigBee, ISA100.11a, WirelessHART, and MiWi specifications, each of which further extends the standard by developing the upper layers, which are not defined in IEEE 802.15.4. Alternatively, 802.15.4 can be used with 6LoWPAN and standard Internet protocols to build a wireless embedded Internet.

Bluetooth Low Energy

Bluetooth low energy—which is also known as Bluetooth 4.0 or Bluetooth Smart—is the version of the Bluetooth Technology specifically designed for the IoT. As its name suggests, this is a power and resource friendly version of the technology, and it's designed to run on low-power devices that typically run for low periods, either harvesting energy or powered from a coin-sized battery.

One of Bluetooth's main strengths is that it has been around for decades, so there are billions of Bluetooth-enabled devices. Also, Bluetooth is a well-established and recognized standard wireless protocol with vast multi-vendor support and interoperability, which makes it an ideal technology for developers. Other advantages are its low peak, average and idle power, which allows devices to run on low-power sources.

The Bluetooth technology makes it ideal for low-power devices such as WSN nodes in the IIoT an M2M edge networks. Bluetooth works over a short range and Bluetooth enabled devices pair with one another in a master/slave relationship in order to communicate. There are several versions of Bluetooth specifications and these are optimized for specific use-cases, such as Basic Rate and Extended Data Rate for speakers and headsets, while Bluetooth Smart will be deployed in IoT smart devices.

However, what makes Bluetooth suitable for WSN nodes is its ability to form ah-hoc networks called *piconets*. A piconet is a grouping of paired Bluetooth devices that can form dynamically as devices enter or leave radio proximity. The piconet consists of two to eight devices that communicate over short distances. (Remember Bluetooth's original purpose as a personal area network (PAN) to wirelessly connect headphones, mobile phones, and other personal electronic devices?) Additionally, Bluetooth ubiquity, miniature format, low cost, and ease of use makes it a very attractive short-range wireless technologies.

ZigBee and ZigBee IP

ZigBee is an open global wireless technology and is specifically designed for use in consumer, commercial, and industrial areas. Therefore, ZigBee is low power and easy to use, and is ubiquitous among IoT networking, sensing, and control applications and devices. ZigBee builds on the standard IEEE 802.15.4, which defines the physical and MAC layers. However, what makes ZigBee different is that it is a superset of IEEE 802.15.4 in so much as it provides application and security layer support. This enables interoperability with products from different manufacturers.

ZigBee works over an area of around 70 meters, but far greater distances can be obtained when relaying communications from one ZigBee node to another in a network.

The main applications for ZigBee 802.15.4 are control and monitoring applications where relatively low levels of data throughput are needed, and low power is a requirement.

ZigBee works in three license-free bands at 2.4GHz, 915MHz for North America, and 868MHz for Europe. Therefore, the ZigBee standard can operate around the globe, although the exact specifications for each of the bands are slightly different. At 2.4GHz, there are 16 channels available, and the maximum data rate is 250Kbps. For 915MHz (North America), there are 10 channels available and the standard supports a maximum data rate of 40Kbps, while at 868MHz (Europe), there is only one channel and this can support data transfer at up to 20Kbps.

ZigBee supports three network topologies—the star, mesh, and cluster tree or hybrid networks. The star network is commonly used, as it is the simplest to deploy. However, the mesh or peer-to-peer network configurations enable high degrees of reliability to be obtained. Messages may be routed across the network using the different stations as relays. There is usually a choice of routes that can be used and this makes the network very robust. If interference is present on one section of a network, another section can be used instead.

The basic ZigBee standard supports 64-bit IEEE addresses as well as 16-bit short addresses. The 64-bit addresses uniquely identify every device in the network similar to the way that devices have a unique IP address. Once a network is set up, the short addresses are used and this enables over 65,000 nodes to be supported.

ZigBee IP

The problem with the original ZigBee standard was that it did not support IP, which was problematic when it came to interfacing ZigBee networks with the real world. However, the ZigBee IP specification is built with a layered architecture and is therefore suitable for any link layer within the 802.15.4 family. ZigBee IP takes advantage of this layered approach and incorporates technologies, such as 6LoWPAN and RPL, that optimize meshing and IP routing for wireless sensor networks. This blend of technologies results in a solution that is well suited to extend IP networks to IEEE 802.15.4-based MAC/PHY technologies.

By incorporating 6LoWPAN into its stack, ZigBee can support IP meshing and routing. Therefore, ZigBee assigns nodes different functions within the IP ZigBee networks, dependent on their position and capabilities.

The nodes can be ZigBee IP coordinator, ZigBee IP routers, and ZigBee IP hosts. Coordinators control the formation and security of networks; this requires an intelligent smart programmable device. IP routers, on the other hand, can be any smart device as they extend the range of networks. Similarly, IP hosts perform specific sensing or control functions at the edge.

ZigBee IP was specifically developed to support IP and the latest ZigBee Smart 2.0 NAN requirements. ZigBee IP supports 6LoWPAN for header compression, IPv6, PANA for authentication, RPL for routing, and TLS and EAP-TLS for security, TCP, and UDP transport protocols. Other Internet applications such as HTTP and mDNS are supported as applications operating over ZigBee IP. However, ZigBee IP nodes and ZigBee standard or PRO nodes cannot reside on the same network. They will require a gateway between them in order to obtain interoperability. ZigBee IP does work seamlessly with other IP-enabled MAC/PHY, such as Ethernet and Wi-Fi.

Z-Wave

Z-Wave is a low-power RF communications technology that is primarily designed for home automation for products such as lamp controllers and sensors. Optimized for reliable and low-latency communication of small data packets with data rates up to 100Kbit/s, it operates in the sub-1GHz band and is impervious to interference from Wi-Fi and other wireless technologies in the 2.4GHz range such as Bluetooth or ZigBee. It supports full mesh networks without the need for a coordinator node and is very scalable, enabling control of up to 232 devices. Z-Wave uses a simpler protocol than some other RF techniques, which can enable faster and simpler development. However, the only maker of chips is Sigma Designs, compared to multiple sources for other wireless technologies such as ZigBee and others. There are many wireless network technologies available for specialized use in various industries.

Wi-Fi Backscatter

A new technology, and probably has more relevance to the consumer Internet but could plausibly have a place in the IIoT, is Wi-Fi backscatter. This technique relies on the fact that radio waves and their energy can be reflected from a Wi-Fi router to power battery free passive devices. The passive device would not only receive the energy it required to operate, but could also reflect some of that energy to another Wi-Fi node in order to send a radio transmission.

One of the biggest issues with the IoT is how all the devices will be powered. Wi-Fi backscatter may actually be the answer. Instead of having all the devices run from button batteries, which will need changed at some point, Wi-Fi backscatter could feasibly provide the power to these devices.

The technology was invented by the University of Washington which developed an ultra-low power tag prototype with an antenna. These tiny tags can be attached to any device and they can talk to Wi-Fi-enabled laptops or smartphones, all the while consuming negligible power.

The tags monitor Wi-Fi signals moving through the air between routers and laptops or smartphones. The tags encode data that they are reflecting, making a small change in the wireless signal. Wi-Fi enabled laptops and smartphones then detect these small changes and can differentiate the signal and receive the data from the tag.

By harvesting power from the ambient RF signals, the passive devices can communicate by reflecting part of that signal to other nodes, albeit with a slight change.

RFID

Another extremely popular wireless technology used in retail, commercial, and industrial IoT is the RFID system. RFID use tags to store electronic information that can be communicated wirelessly via electromagnetic fields. Dependent on the system, tags can be either passive or active and this will determine the range from which the tag and the RFID reader can operate. For example, some tags will use the energy from the reader's own interrogating radio wave and simply act as a passive transponder. Other passive tags can be powered through electromagnetic induction, which is generated when they come in close proximity to the reader. These types of tags have very short operating ranges as they require being close to the electromagnetic source. On the other hand, some tags are battery powered and transmit actively and these can be situated hundreds of meters from the reader.

RFID technology is used in many industries to identify and track inventory, people, objects, and animals due to the tag's versatility and ability to be attached to just about anything. RFID is also used in contactless payment systems using cards or even smartphones to place in close proximity to the RFID reader. However, RFID does not always require such close contact; in some cases even the briefest contact at a distance is all that is required. An example of this is the timing of sports cars lapping a track. Even at those high speeds, RFID works efficiently and reliably and produces accurate timing.

Another advantage of RFID is that the tags do not need line of sight or even need to be visible, so they can be easily concealed in packaging and products. RFID tags can be read simultaneously by a reader if they are in range, which is a big advantage over barcodes, which are read one at a time. Hundreds of RFIDs can be read at once.

Miniaturization has greatly enhanced the use of RFID, as now tags can be microscopic in size. Hitachi has so far produced the smallest miniaturized RFID tag at 0.05mm x 0.05 mm and these dust-sized tags can hold a 38-digit number on a 128-bit ROM.

NFC

Near Field Communication (NFC) evolved from RFID technology and is now a common method of short-range wireless communication use in contactless payment systems. The way that NFC works is that a chip embedded in a bank-card, for example, acts as one end of a wireless link when it becomes energized by the presence of another NFC chip. As a result, small amounts of data can then be passed between the two chips. With NFC, no device pairing is necessary and the range is limited to a few centimeters. NFC has become popular with smartphone application developers to create payment systems or methods to transfer contacts between phones simply by placing them close together.

Thread

A very new IP-based IPv6 networking protocol aimed at the home automation environment is called Thread. Based on 6LoWPAN, Thread is not an IoT applications protocol like Bluetooth or ZigBee. However, from an application point of view, it is primarily designed as a complement to Wi-Fi as it recognizes that while Wi-Fi is good for many consumer devices, it has limitations for use in a home automation setup.

Launched in mid-2014 by the ThreadGroup, the royalty-free protocol is based on various standards, including IEEE802.15.4 (as the wireless air-interface protocol), IPv6, and 6LoWPAN, and offers a resilient IP-based solution for the IoT. Designed to work on existing IEEE802.15.4 wireless silicon from chip vendors such as Freescale and Silicon Labs, Thread supports a mesh network using IEEE802.15.4 radio transceivers and is capable of handling up to 250 nodes with high levels of authentication and encryption. A relatively simple software upgrade should allow users to run thread on existing IEEE802.15.4-enabled devices.

6LoWPAN

Low-power radio devices must transmit their data in a timely and efficient manner and to do that there needs to be improvements in the communication protocols. Existing IP protocols have far too much packet overhead for most embedded sensors, which transmit only tiny quantities of data. The ratio of overhead to data is unacceptable, so a new more efficient communication protocol that was still IP compatible was required, called 6LoWPAN.

6LoWPAN is short for IPv6 over low-power personal area networks. 6LoWPAN provides encapsulation and header compression for IPv6 packets, which provides briefer transmission times.

Rather than being an IoT application protocol technology like Bluetooth or ZigBee, 6LoWPAN is a network protocol that defines encapsulation and header compression mechanisms. The standard has the freedom of frequency band and physical layer and can be used across multiple communications platforms, including Ethernet, Wi-Fi, 802.15.4, and sub-1GHz ISM. A key attribute is the IPv6 stack, which has been a very important introduction in recent years to enable the IoT.

IPv6 offers approximately 5×1028 addresses for every person in the world, enabling any embedded object or device in the world to have its own unique IP address and connect to the Internet. Especially designed for home or building automation, for example, IPv6 provides a basic transport mechanism to produce complex control systems and to communicate with devices in a cost-effective manner via a low-power wireless network.

Designed to send IPv6 packets over IEEE802.15.4-based networks and implement open IP standards including TCP, UDP, HTTP, COAP, MQTT, and web sockets, the standard offers end-to-end addressable nodes. This allows a router to connect the network to IP. 6LoWPAN is a mesh network that is robust, scalable, and self-healing. Mesh router devices can route data destined for other devices, while hosts are able to sleep for long periods of time.

RPL

Routing issues are very challenging for low-power networks such as 6LoWPAN. This is due to devices operating over poor lossy radio links, and the situation comes about due to the low power available to battery-supplied nodes. Compounding the problem is the multi-hop mesh topologies and the frequent topology changes. This is especially true if mobility is another design consideration.

One solution to operating over lossy networks is RPL, which can support a wide variety of different link layers, including ones that are constrained and potentially lossy. Designers typically utilize RPL in conjunction with host or router devices with very limited resources, as in building/home automation, industrial environments, and urban applications.

RPL efficiently and quickly builds up network routing knowledge, which it can distribute among its connected nodes. The RPL nodes of the network are typically connected through multi-hop paths to a small set of root devices. The root devices are responsible for route information collection. Each route device creates a Destination Oriented Directed Acyclic Graph (DODAG), which is a visualization of the surrounding network, which takes into consideration the link costs, node attributes, and status information so that it can plot the best path to each node.

RPL can encompass different kinds of traffic and signaling information exchanged between host. It supports Multipoint-to-Point (MP2P), Point-to-Multipoint (P2MP), and Point-to-Point (P2P) traffic.

Proximity Network Communication Protocols

When considering a communication protocol for the proximity and access network, you have to consider the diversity and capabilities of the existing devices and communication lower layer protocols. As you learned in the previous section, not all protocols support IP, so you have to consider these protocols when attempting to find a common data and network layer protocol.

Internetwork Protocol (IPv4)

IP comes in two distinct flavors—IPv4 and IPv6. IPv4 is considered to be at the end of its life and we are being urged to transition to IPv6. However, this has been going on now for well over a decade with each year being trumpeted as the year for mass IPv6 deployments. Each year passes with no further progress. This is because business, commerce, and industry see little advantage, but potentially huge disruption and costs in migrating to IPv6. Therefore, change has been much slower than expected. The Industrial Internet may change that as IPv6 is seen as a key technology enabler for the IIoT.

For now though, IPv4 is vastly more common around the globe. Since technologists have found ways to overcome many of IPv4's perceived shortcomings, it is likely to be around for some time to come. Therefore, within the context of the IIoT we have to consider both variants as undoubtedly IPv4 will be around for at the very least another decade.

As a result, we will briefly take a high-level look at both protocols, including how they work, what you need to know about working with each, and how to have them coexist within the network of things.

The obvious difference between the versions is the address format. An IP address is the unique identifier that the Internet Protocol (IP) assigns to each node or host in a network. A host is anything that requires connecting and communicating with other hosts over the network. Hosts can be computers (servers and PCs), network printers, smartphones, tablets, and even TVs and fridges. A network is a group of two or more hosts that connect using a common protocol (language) of communication. Today, the most popular and widespread method of connecting hosts is by Ethernet or Wi-Fi, and the most popular protocol is IP.

An IP address is actually a 32-bit binary address as computers and the processors embedded within other hosts talk in binary. However to be human readable they are referred to in dot-decimal notation, which consists of four decimal numbers, separated by dots, each between 0 and 255. Therefore, the 32-bit binary address is split into four octets (8-bit) and maps to four decimal numbers separated by dots.

IP ADDRESS			
172	16	1	254
10101100	00010000	00000001	11111110
32-bit binary address –> 4 x 8 bits binary			

All IPv4 addresses conform to this four-byte format.

IP has been around a long time and its method of addressing has evolved as circumstance dictated. During the pre-Internet days, IP addresses were used freely and without any real consensus as to what part was the network and what part was for hosts. Clearly for the protocol to succeed, there had to be an agreed structure so that anyone receiving an address packet could ascertain what network it belonged to and what its host identifier was. The resulting standard was the IP classes A, B, C, and D, with a fifth E reserved.

Class	Start Address	End Address	Net Bits	Hosts Bits	No. of Subnet	Non of Hosts	Leading ID Bits
A	0.0.0.0	127.255.255.255	8	24	2^7	2^{24}	0
B	128.0.0.0	191.255.255.255	16	16	2^{14}	2^{16}	01
C	192.0.0.0	223.255.255.255	24	8	2^{22}	2^{14}	11
D	224.0.0.0	239.255.255.255	Multicast Address Space				111

Class A addresses where handed out to huge or technology companies such as IBM, Microsoft, and Apple. The class B address space was designated to large companies and Universities while class C addresses were designated for small companies. This was the standard policy for some time, until in the early 90s came the rapid expansion of the Internet. In response to the huge demand for IP addresses, and in much smaller segments, IANA introduced CIDR (Classless Inter Domain Routing) in 1993. CIDR is based on variable-length subnet masking (VLSM), which does not have fixed network and host segments. Instead it allows the boundary to be flexible. The advantage of having a flexible mask is that it allows the network designer to adjust the mask to create any size of subnet.

VLSM was the basis of flexible-length subnet masking and this made the notion of fixed classes redundant. VLSM works on the principle of bit borrowing by telling the subnet mask to land on any bit boundary. No longer were the legacy class boundaries of 8, 16, and 24 relevant; now a boundary could sit at 17 or 28. As a result, the flexible subnet mask was introduced that represented by either a shorthand /(no of bits) such as /8 or as the traditional dot-decimal notation.

The growth of the Internet also determined change in other respects. Prior to the Internet, IP addresses were assigned to whomever asked in class blocks. There was no real demand and therefore no shortage of IP addresses. In hindsight it was a dreadfully wasteful policy. However with the explosive surge in demand for IP addresses in the mid 1990s, it became clear that IPv4, despite its seemingly vast address space of 2^32 or over four billion addresses, was unsustainable. In order to address the problem, private IP addresses were introduced.

Private IP addresses were a tactical move to mitigate the problem of chronic IP address shortage. Three blocks of addresses, taken from the original IP address classes A, B and C, were designated as private and they were reserved only for use within the boundaries of private networks.

Address Range	Address Block	Largest CIDR Block
10.0.0.0 – 10.255.255.255	24-bit	/8
172.16.0.0 – 172.16.255.255	20-bit	/12
192.168.0.0 – 172.16.255.255	16-bit	/16

IANA designated these ranges of IP addresses for use by anyone to address their own private networks. This decision was based on the assumption that very few computers in use in private businesses or government actually needed to connect to the Internet. This was indeed the case and network administrators slowly began to adopt private addressing within their business networks. Private addressing became more readily acceptable as reports of security issues with the Internet began to emerge. However, there was a newly emerging technology, which made private addressing not just grudgingly acceptable to businesses but downright desirable.

Network Address Translation

NAT, or network address translation, is a technology implemented in routers and firewalls that translates private (illegal/meaningless on the Internet) addresses to real public IP addresses. This meant that computers configured with a private address could now access the Internet via the router or firewall.

Administrators could manage these single points of exit using access lists and the real beauty was it was one way. A computer with a private address could initiate communications with a web site on the Internet, establish a TCP session, and communicate with another computer on the Internet, but the computer was safe from being directly accessed by a system on the Internet. This was because NAT would prevent an Internet host from accessing (initiating the TCP handshake/session) a private address host on the private network.

Private addressing and NAT went a long way toward prolonging the life expectancy of IPv4, as did virtual hosting of web sites and Internet cloud services. However, it was not to last, as other disruptive products sent demand for IP rocketing. ADSL and smartphone technology, along with tablets, sent IPv4 once again to the brink of exhaustion. Furthermore, business interest in the successor technology IPv6 has been cool, with every year passing by with more and more excuses for not migrating across.

IPv6 solves all the problems of IPv4 through a limitless supply of addresses, automatic addressing, and no more subnetting headaches. IPv6 is the future—there is no one that will contest that—but adoption has been painfully slow. Perhaps that should be interpreted as IPv4's greatest compliment.

IPv6

So why do we need a new IP version, if IPv4 is so popular?

The first key driver for a revised IP version is that the Internet has grown exponentially and the global demand for IPv4 address space has saturated the available address pool. The Internet of Things has also provided the necessity for a larger address pool as it requires a protocol that has an address space large enough to cope with the demands of the potentially vast amounts of "things" that will be eventually connected to the Internet. IPv6 solves this problem by extending the addressable address space from 32 to 128 bits.

Another reason is that IPv4 was designed pre-Internet, or rather in the Internet's infancy when only a few universities and government establishments were connected. Consequently, IPv4 lacks many features that are considered necessities on the modern Internet, for example, IPv4 on its own does not provide any security features. With native IPv4, data has to be encrypted using some other security encryption application, such as SSL/TLS, before being transported across the Internet. In contrast, IPv6 has IPSec as a requirement so every packet is encrypted without the need for another security mechanism.

IPv4 has an inefficient header-to-payload ratio, and the header is unnecessarily complicated. IPv6 simplifies the header by removing many of the IPv4 options, or by moving them to extensions at the end of the packet. An IPv6 header is therefore only double the size of an IPv4 header despite the packet being four times the size. The increased header efficiency and simplicity enables faster routing despite the larger packet size.

In a similar way, IPv4 has no efficient way for providing data with strict quality of service, as prioritization in IPv4 is based on priority classesan drop eligibility. IPv4 achieves this via the six bits DSCP (Differential Service Code Point) and the bits ECN (Explicit Congestion Notification) to provide some basic granularity between each class. IPv6, on the other hand, has a built-in quality of service across all IPv6 speaking nodes. IPv6 achieves this using traffic class (8 bits) and flow labels (20 bits), which are used to tell the underlying routers how to efficiently process and route the packet. The traffic class uses the same 6-bit DSCP (Differential Service Code Point) and 2-bit ECN (Explicit Congestion Notification) as IPv4; however, the 20-bit flow label is set by the source to identify a session flow, which is used in streaming applications. The improvement in IPv6 headers means that QoS is maintained even across the public Internet, which is impossible with IPv4.

Another problem with IPv4 is the administration burden of maintaining the address configuration of IP enabled clients, as these can be configured manually or they need some address configuration mechanism. In order to address this issue, IPv6 configures its own unique global address, as well and its own local link address, without any administrative burden.

IPv6 also has many other notable features that provide benefits over IPv4, such as its extensibility. It can support far more options and future extensions than IPv4. IPv6 also provides greater mobility, which is something in great demand today, as it provides the ability for mobile phones to roam to different geographical areas and still maintain their own IP addresses.

IPv6 also has streamlined the packet and delivery methods by doing away with the very bandwidth-hungry broadcast method, preferring unicast, multicast, and anycast delivery mechanisms.

Address Types

IPv6 has 128 bits of address space split into two 64-bit sections with the second half of the address the last 64 bits being the interface address. To make the interface address unique, the IPv6 protocol uses the network interface's 48-bit MAC address as the unique identifier. In order to maintain uniqueness, the host takes its own MAC address and splits it into two parts. It then inserts the byte's FF FE in between the two segments, constructing a unique 64-bit network interface address.

Where IPv6 differs significantly from the IPv4 addressing concepts is that IPv6 is designed to have multiple addresses per interface. For example, it has a global address, which is equivalent to an IPv4 public address. It also has a link local address, which IPv6 uses to communicate with other hosts over a shared link or subnet. Finally, there is the unique local address, which is equivalent to

Chapter 5 | Designing Industrial Internet Systems

an IPv4 private address. The three most significant bits can distinguish these three address types from one another. The global address always has the three bits set to 001. Link local is set to FD, and unique local is set to FE.

IPv6 Subnets

One of the most dreaded features of IPv4 was the concept of the variable length subnet mask, which was used to subnet the address space into network and host bits. With IPv6, things are much simpler, as the 64 least significant bits of the 128-bit address (the last 64 bits) always indicate the host address. The subnet is a separate block of bits 16 bits in length that indicates the subnet, which provides 65,000 possibilities. That's more than enough.

Figure 5-4 shows how the routing address (48 bits), the subnet (16 bits), and the host address (64 bits) are derived from the 128 IPv6 address.

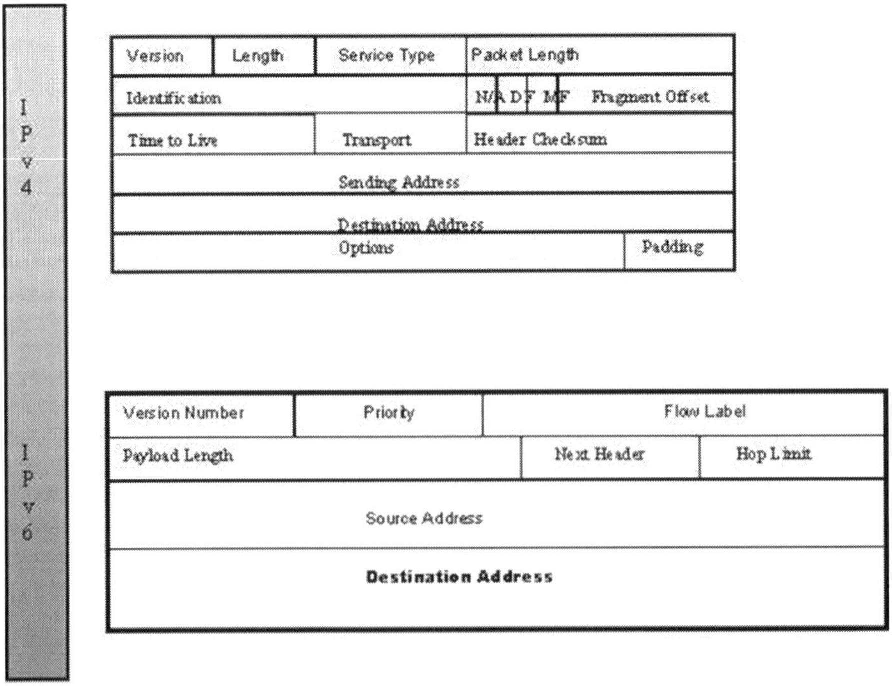

IPv4			
Version	Length	Service Type	Packet Length
Identification			N/A DF MF Fragment Offset
Time to Live		Transport	Header Checksum
Sending Address			
Destination Address			
Options			Padding

IPv6			
Version Number	Priority		Flow Label
Payload Length		Next Header	Hop Limit
Source Address			
Destination Address			

Figure 5-4. Ipv6 header

IPv6 for the IIoT

In the previous section, we considered the differences between IPv4 and IPv6; however, what specific features of IPv6 make it ideally suited to the IIoT?

The most obvious reason for using IPv6 rather than IPv4 is the vast availability of global IP addresses. The sheer scalability of IPv6 is staggering. It provides 2,128 unique addresses, which represents 3.4 × 1038 addresses, a truly huge number. However, it is not just that these addresses will be inexhaustible, it is also that IPv6 negates the requirements of NAT (network address translation). NAT breaks the true Internet model because although hosts in a private network can access web servers on the Internet, the reverse in not true and NAT hosts are unobtainable from the Internet. This is sometimes looked as being a major benefit of NAT in a security and Enterprise context. However, in an IIoT context, NAT is not desirable, as having global access to "things" is desirable.

Furthermore, IPv6 is suited to the Internet of Things as there is a compressed version of IPv6 named 6LoWPAN. It is a simple and efficient mechanism to shorten the IPv6 address size for constrained devices, while border routers can translate those compressed addresses into regular IPv6 addresses.

In addition, the problem regarding constrained devices hosting an IPv6 stack has been addressed via tiny stacks, such as Contiki, that require no more than 11.5KB.

Gateways

Having examined the myriad of communication technologies—both wired and wireless—which enable devices (things) to be interconnected within the proximity network, it is necessary to envisage how we connect all these variants of protocols to interface with an IP backbone. Initially, the access network, which is the communication channel that connects the edge devices to the application servers and controllers, would be a trivial process. However, it is far from that as we have to translate non-IP protocols such as Rs232, 4-20ma, the wireless protocols such as ZigBee, Bluetooth, and even wired real-time protocols such as Profinet and field bus to interface with a standard IP interface. After all, these protocols have profound electrical and engineering differences.

The way that this is accomplished is via gateways, and these are devices usually strategically placed to translate and convert the electrical characteristics of one protocol to another. A simple example is a serial RS232-to-USB converter. Of course, there are many other technologies, such as field bus to Ethernet and 4-20ms to Ethernet, that we can encapsulate and tunnel through an Ethernet network. Sometimes, we may be only tunneling the native payload

Chapter 5 | Designing Industrial Internet Systems

through Ethernet to deliver to another field bus node. In that case, encapsulation is performed at the ingress to the Ethernet network and decapsulation takes place at the egress, thereby delivering the datagram in the same format. Furthermore, we can tunnel wireless technologies such as ZigBee and Bluetooth, and all the others, in a similar manner in order to pass their payload to a similar application or cloud application.

Gateways are devices that enable us to transition one protocol to another, in the case of the Industrial Internet it is typically between a native protocol and Ethernet. To show how this works, look at Figures 5-5 and 5-6.

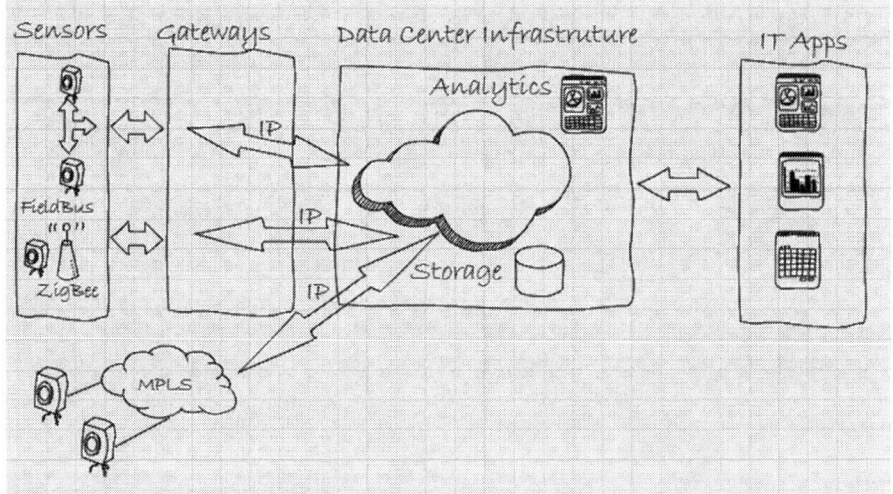

Figure 5-5. Heterogeneous networks

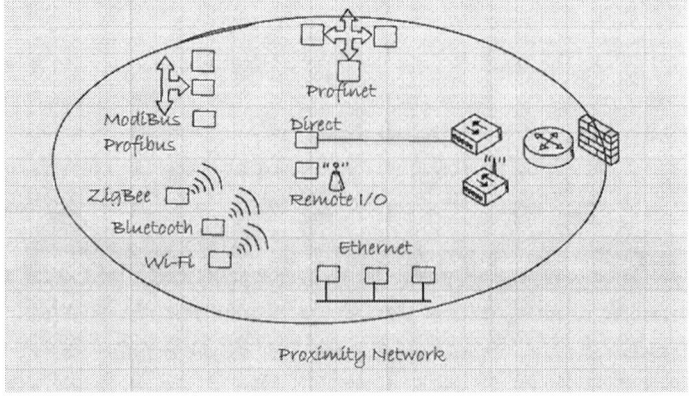

Figure 5-6. Diverse protocols in the proximity network

Industry 4.0

Figure 5-5 shows field bus nodes communicating over an Ethernet Backbone. In Figure 5-6, you can see ZigBee and Bluetooth devices interfacing with an IP access network running IP. Then, Ethernet on copper transforms to an MPLS network running on optical fiber to establish a private line.

The point is that gateways allow us to integrate diverse technologies and protocols onto a common network. This is hugely important when building interconnecting networks. You can build interface complex and diverse heterogeneous networks that look like Figure 5-5.

The main takeaway here is that it doesn't matter which technology we use in the proximity network to connect our devices; we can still interface through gateways to a common IP infrastructure. Look again at the original simplistic model in Figure 5-6.

We can see that we are not restricted to IP in the proximity network so long as we have a gateway that can handle the transition from the natural format— for example, RS232 to IP/Ethernet.

The big question that then arises, is if there are gateways available to handle this electrical and data format transition between such diverse technologies. Well, yes, there are, and we can demonstrate this in Figure 5-6, where we see many diverse technologies bridging or translating across a gateway device.

It may all sound easy, but here is the problem. Consider the automobile field bus technology CAN. This technology was developed to work at layer 2 of the OSI model, so that every device installed win the car can communicate. However, an application without the car cannot communicate, as there is no application protocol established at layer 7. Consequently, we need an open version of CAN that will allow cars from different manufacturers to communicate by establishing a common layer 7 protocol—hence OpenCAN and DeviceNet. These two cars, or rather their devices, can communicate as they share a common layer 7 application design.

Similarly with IP and Ethernet, all devices that share the same standard can communicate as they share the same common layer 1-4 layer protocols. However, they can communicate but not necessarily understand each other as they do not share the same layer-7 application language. The way that commercial IP/Ethernet gets around this problem is by using standard software protocols such as HTTP, FTP, and SNMP in order to form a common language. However, with Industrial Ethernet it does not require these high-level application protocols. What it need is a real-time protocol to handle data as message exchanges.

However, how do these gateways convert traffic from one network to another? The answer is that they have several techniques; routers for example come with several different interfaces and they inherently have the intelligence and logic to drive and route traffic between them. Therefore, a standard industrial router can accept traffic on an Ethernet port and route that traffic out of a WAN DSL port—in the correct frame format—without any problem. Similarly, it will accept Ethernet traffic through one port and transform the frames to merge into an ATM cell stream and vice versa.

There are also media converters available on the market that can convert RS232 to USB or even optical to electrical (fiber to copper) and vice versa. Another technique that may sound like a media converter is the way a device server converts serial RS232 to Ethernet. Under the hood, the device server simply encapsulates the serial data into an Ethernet frame for transmission over the LAN. Then, at the recipient, the packet is de-encapsulated revealing the serial data. Timing is imperative though, and serial to Ethernet can be tricky as serial operates only at layer 1. Because serial communications do not have an application layer protocol and no application port number, this can make data linking to an application difficult and it will have to be done programmatically.

Industrial Gateways

An industrial gateway works by linking traffic flowing from one system to another. It works despite the fact that they might not use the same communication protocol or even similar data formatting structures. Gateways, like modern routers, use the entire OSI model from Layer 1 up to Layer 7. This allows a gateway to perform true data translation and processing on the traffic flowing between the two different networks. Gateways, unlike routers, are cheaper, as they perform only protocol and frame translation and are built and ordered to perform specific translations. However, when we use IP, whether in the proximity or the access network, we require routers to manage the flow of IP traffic between subnets or VLANs.

CHAPTER 6

Examining the Access Network Technology and Protocols

In the previous chapter, we had to consider many diverse technologies and protocols that are in use today in Industrial Internet scenarios. However, thankfully within the access network segment we are on far more traditional network territory with long-established and dominant technologies and protocols. Therefore, in this chapter we will look briefly at the communications technologies that are likely to be implemented in an Industrial Internet deployment and the corresponding protocols and applications that are deployed as middleware to facilitate the specific requirements for some Industrial Internet use-cases.

The Access Network

The access network in the IIC's architecture of the Industrial Internet is purely a transitional transport network with the purpose of interconnecting the diverse networks and protocols, which exist in the complex proximity network with the standard IP networks found in enterprise and business environments. Therefore, the technologies and protocols that we will examine are used to interconnect the domains and transport data back and forth and these are the same technologies and protocols commonly deployed in the enterprise.

The sole purpose of the access network is to aggregate data and to backhaul it from the device gateway to the system and applications within the operational and management domain. As this domain is purely IT, it follows that it will use a traditional IT architecture, based on IP and Ethernet. Therefore, the principle task for the components in the access network is to transport IP over Ethernet as efficiently as possible, and that will require gigabit Ethernet trunks over copper, optical fiber, or even wireless and high-speed Ethernet switches.

Ethernet

So what is an Ethernet frame? A data packet on an Ethernet link or LAN is actually an Ethernet packet and it has an Ethernet frame as the payload. An Ethernet frame is represented in Figure 6-1.

Layer	Preamble	Start of Frame Delimiter	MAC Destination Address	MAC Source Address	802.1Q Tag	Ether Type	Payload	Frame check	Inter Packet gap
	7 octets	1 octet	6 octet	6 octet	4 octet	2 octets	46 – 1500	4 octets	12 octets
2		Layer 2 Ethernet Frame (64k -1518k)
1	Layer 1 Ethernet Packet (1530 octets)

Figure 6-1. Ethernet frame

An Ethernet frame has the very first field as the destination MAC address and the second field is the source MAC address. However, there is another field of interest that is the frame following the MAC source address. This frame is called the 802.1q tag.

VLANs

As traffic will be aggregated from multiple edge networks, we will have to deploy some form of tagging in order to track which traffic belongs to which edge network. By tagging the traffic, we can effectively segregate each individual stream and provide a virtualized network. This technique in Ethernet is called a VLAN or Virtual LAN. A VLAN provides basic levels of segregation, security, and efficiency as the size and range of broadcast domains are reduced—although the number of broadcast domains actually increases—and this reduces the amount of broadcast traffic and the corresponding host responses, considerably.

VLANs are a popular way for a network administrator to segment a layer-2 switched network by configuring the local Ethernet switches with VLAN IDs and then placing the selected ports into the designated VLANS. This is great for local administration of a single switch. However, what if the requirement is for the VLAN to span the entire network of 1,000 switches? Then the solution requires VLAN trunking and a VTP management domain.

When configuring VLANs on an Ethernet switch, the ports can be either access ports, which directly connect to a host, or trunk ports, which connect to another switch. There is another difference though—an access port can only accept and forward packets from one VLAN, whereas a trunk port accepts and forwards packets from two or more VLANs.

A trunk port is a point-to-point link to another switch or router that forwards traffic to several destinations. Trunk ports therefore carry information from several VLANs across the same physical link, which allows VLANs to span the network. However, for a packet to leave its local switch where the VLANs are known—the local VLAN database—there has to be some way in which the packet's VLANs membership can be determined by another switch.

The way an Ethernet network handles this is through tagging the individual packets with a VLAN ID. Every packet entering the VLAN port on the local switch is tagged with a VLAN identifier. Therefore, every packet that is in a VLAN will carry with it a special tag declaring its local VLAN ID that other switches on the network can inspect and forward accordingly. The switch does this by placing a tag in one of the fields in the Ethernet frame.

This VLAN identifier is how an Ethernet frame can be marked to be different from other frames. The VLAN tag defines its VLAN membership. Now the network switch, when receiving a packet, can look inside and check the frames value for not just the source and destination addresses but also the VLAN ID. Additionally, administrators can now group hosts from different network segments many switches away from one another into the same virtual LAN as if they were sitting on the same local switch.

When a network grows, maintaining and administering VLAN architecture becomes a burden as VLAN configuration information has to be configured and updated across all the switches in the network. Administrators then need to use VTP (VLAN Trunking Protocol) or something similar in order to maintain consistency across the network. VTP assists in maintaining, renaming, updating, creating, and deleting VLANs on a network-wide basis. VTP minimizes configuration inconsistencies such as incorrect names or VLAN types.

IP Routing

The alternatives to high-speed Ethernet switching (layer 2) are to do IP routing (layer 3) between the various edge networks and the operational and management platforms. Routing is generally much slower than Ethernet switching, as routing is predominantly a software function with many lookups of IP routing and forwarding tables.

IP routing does provide for subnetting of the IP address space, which provides similar functions as the Ethernet VLANs discussed earlier, in that they provide segregation and security, and they break up large broadcast domains. Subnets are not required with IPv6, which is one of its great advantages.

Routing, if pure speed of delivery is not the ultimate goal, has many distinct advantages, such as it provides more granular control over traffic flows, traffic management, and quality of service (QoS).

Furthermore, in this instance, we have assumed that the edge networks and the operational and management platforms are in the same location or at least within a layer-2 switching LAN, as is usually the case in commercial and manufacturing. However, what if these edge networks are remote, perhaps 100s of mile away, geographically dispersed around the country, or even the globe? How would we construct our access network then?

Access Networks Connecting Remote Edge Networks

Routing over WAN links is the most obvious choice and we will discuss those options later in the WAN section. However, we can still in most cases use fast Ethernet switching. For example, we can connect remote office, production sites, or factories using a service provider's MPLS network (multiprotocol label switching). The service provider can design layer-2 VPNs across their MPLS network to provide what is effectively a layer-2 overlay, which will connect all the remote locations as if they were on the same layer-2 segments. This service can be termed as a virtual private wire or LAN. Similarly, dark fiber—other's spare capacity—can be utilized in metro and urban areas to interconnect data centers using optical long distance point-to-point Ethernet interfaces.

Carrier Ethernet

Another industrial variant of Ethernet is Carrier Ethernet and it is commonly used in industrial scenarios for interconnecting switching rooms and substations. An example is the way mobile operators interconnect their switch rooms at their remote and geographically dispersed tower locations. By using carrier Ethernet or MPLS, mobile operators can switch backhaul traffic from the towers as reliably and deterministically as they once did with synchronous technologies such as ATM and TDM.

Whichever method we chose, whether routing or switching, the end goal is the same—to deliver the traffic from the edge networks to the operational and management platforms. Once again there are many ways to achieve the same goal; it all really depends on the specific requirements as to which is the better solution.

Profinet

Switching and routing performance may be sufficient for most enterprises and commercial activities and use-cases. However, in manufacturing, the non-deterministic performance of these Ethernet and IP solutions fall short of the requirements. Industrial solutions especially in manufacturing requires highly robust, reliable, and real-time solutions. The problem is that IP is just not quick enough in that it introduces too much latency (delay) or deterministic, in so much as it introduces too much jitter.

Industrial Ethernet covers many of these issues and a leading open standard for Industrial Ethernet in manufacturing is Profinet.

In an early chapter, we discussed briefly the differences between standard Ethernet and Industrial Ethernet, so we will not revisit those areas. However, the reason we are discussing Profinet in this section is to show why and how Profinet—as a leading Industrial open standard—resolves the issues of latency and jitter that bedevils standard switching and routing.

Profinet is 100% Ethernet compatible and adheres to IEEE standards so can be deployed in a flexible line, ring, or star topology using copper and fiber-optic cable solutions. It also enables wireless communication with WLAN and Bluetooth.

However, Profinet also includes intelligent diagnostic integrated tools for field devices and networks. These tools provide important information regarding the status and health of devices and the network, including a display of the network topology. Profinet delivers this diagnostic information through acyclic diagnostic data transmissions, which allows redundant high available equipment to react automatically to any predicted failure, enhancing the stability and availability of the manufacturing plant.

However, what is of interest to OT network designers is how Profinet handles data transmission. Here we see that Profinet handles all communications over the same cable. The communications can range from simple control tasks to highly demanding motion-control applications. Indeed, for high-precision closed-loop control tasks, deterministic and isochronous transmission of time critical data is handled with a jitter of less than 1μs (see Figure 6-2).

Chapter 6 | Examining the Access Network Technology and Protocols

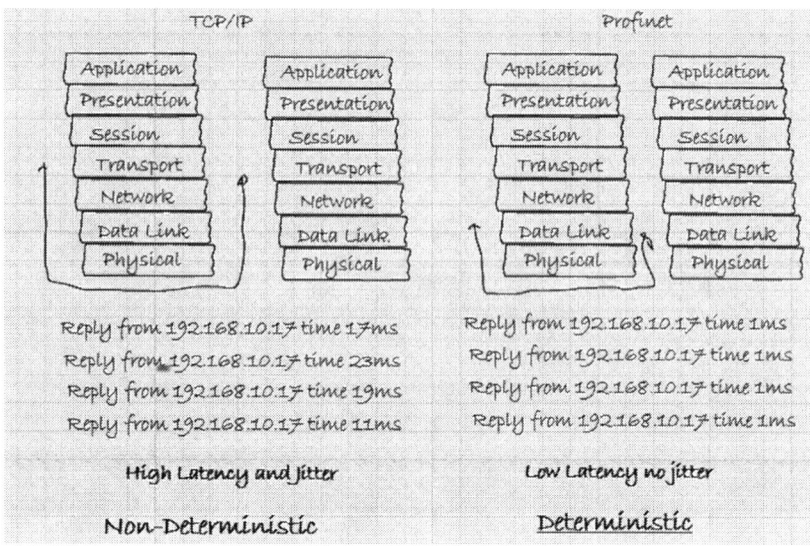

Figure 6-2. Profinet's deterministic nature

Synchronous Real-Time

Profinet achieves its deterministic performance because, like all Ethernet switches, it uses its MAC address to communicate with other Ethernet devices over the local LAN. By not using an IP address, Profinet can reduce latency and jitter considerably. However, Profinet also provides deterministic (very little jitter) by the way it manages data flows. What Profinet does is split data transmissions into three types:

- Cyclic I/O data transmitted in real time
- Acyclic data transmissions used for parameter data and diagnostic information, which is not in real time
- Alarms, which is another real-time transmission channel for urgent maintenance required status alarms

By using dedicated cyclic I/O data channels for input and output, a Profinet I/O controller can set sub-channels and determine the clocked phase of each sub-channel within a range of 250μs to 512ms. Therefore, the I/O controller can control the transmission latency and amount of jitter by setting and monitoring the cyclic I/O data channels to, for example, 1ms.

Therefore, Profinet controllers are capable of handling applications with the most stringent deterministic demands. They can minimize the jitter through synchronous cyclic I/O data transmissions. However, for all the devices to handle synchronous data transmissions with a maximum deviancy of 1μs, they must have a shared clock.

Examining the Middleware Transport Protocols

With regard to the connectivity between devices in the proximity edge network and communication between the operational and management domain, we have to bear in mind that diverse protocols and layer-1 physical electrical characteristics will differ. As an example, we might well have, in a manufacturing plant, many devices connected by diverse technologies such as Bluetooth, ZigBee, or even Ethernet. The problem is to integrate those protocols so that it is transparent to the systems. We discussed gateways earlier, but they introduce latency, jitter, and potential packet loss.

The reason for this is that not all protocols follow the same goal. For example TCP/IP, which is a suite of protocols developed a long time ago to enable reliable transport of data over very unreliable and lossy networks, is wonderful for exactly that, transporting data over unreliable networks, which were common back in the 80s.

Chapter 7 | Examining the Middleware Transport Protocols

However, as network technology has advanced over the years, networks have become far more reliable so TCP/IP, which requires a connection between communicating partners is no longer necessary, within the context of local transmission. For example, if you are communicating across a LAN on Ethernet, the capabilities of TCP (transport control protocol) to deliver reliability, lost packet detection, resend of lost packets, and packet resequencing—packets may turn up out of order—are no longer required.

However, that does not mean that IP alone can handle the requirements of Industrial Internet. Not even on a LAN or crossing a private line/LAN on a carriers MPLS network can we assume zero packet loss. As a result, TCP/IP is still commonly used over WAN links and potentially lossy networks.

TCP/IP

TCP/IP is a suite of protocols that enables IP data to traverse unreliable networks such as traditional dial-up and long distance serial links with a measure of reliability. TCP/IP provides IP with the functionality to detect lost packets—by sequencing each packet—as well as to enable the recipient to reorder packet streams according to their sequence numbers.

The problem is that TCP is connection orientated, for example, two communicating hosts require establishing a connection known as a *session* in order to transmit packets. The session between hosts is established through a three-way handshake, after which both hosts can communicate freely with each other.

Figure 7-1 shows how the three-way handshake is established and maintained to keep the session between the pair active.

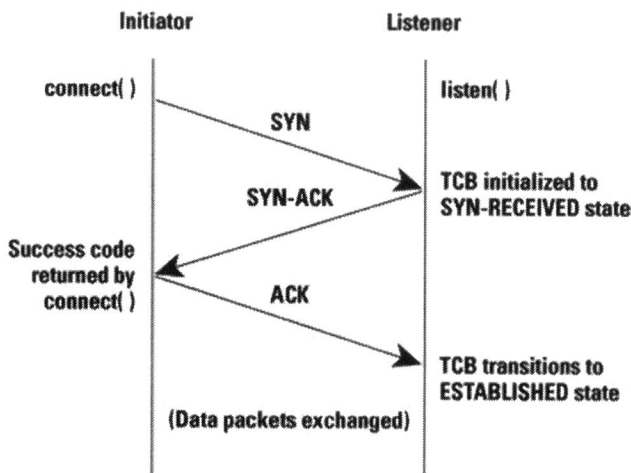

Figure 7-1. TCP three-way handshake

TCP/IP was developed decades ago yet has maintained its role as the dominant communication protocol to this day. This is due to its fantastic ability to deliver communication of data reliably across even dodgy media. The problem, though, is that modern communications require real-time transportation. For example, resending packets or reordering packets received out of order is perfect for legacy protocols such as FTP (File Transfer Protocol), HTTP downloads and the like, but is useless for video streaming or VoIP.

After all, with real-time applications such as a VoIP conversation, what is the point of requesting a lost voice or video packet to be resent? It will do more harm than good, similarly with streaming video, lost or out of sequence packets are treated as lost and as long as there are not too many of them, it will be transparent to the viewer.

However, there is another problem with TCP/IP and even with IP itself, and that is it is inherently slow. To understand this, we need to understand what actually happens when data is transmitted from one host to another. An example is a web client sending a URL request to a web server, and this is simple enough to explain. The client looks first to resolve the URL to an IP address via DNS (dynamic name server) and then uses the registered IP address that corresponds to the URL to contact the server.

The server will receive the request on the wire and recognize its unique MAC IP address and application-specific port number and will send the request up the layers to the application layer that can make sense of the request. The recipient's application layer will handle the received request and send the response back down through the OSI layers in order to package and frame the data, ready to be transmitted to the source host. An example is shown in Figure 7-1.

The problem is that TCP/IP (layers 1-4) and IP (layer 3) require each host to establish a connection, and that each layer be able to process the data, whereas Ethernet only requires data to be processed at layer 2. As a result, TCP/IP can be very slow compared to other connectionless, unreliable data transport technologies, such as UDP.

UDP

UDP stands for Unreliable Data Protocol, and that is what it actually is. It is connectionless and just transmits data onto the media (wire, optical, or wireless) on a "hope-it-gets-there" basis.

When we discuss Industrial Internet protocols, UDP is often suggested as the low overhead efficient protocol for IIoT, due to its low ratio of payload to header overhead. This is true, however, the name itself should warn you of the problem and it is unreliable. UDP has no mechanisms to check out of sequence packets, or lost packets, it just fires and forgets. If it gets there great; if not, well who cares.

Therefore, do not consider UDP to be an alternative for time-critical and mission-critical systems just because it is quicker and more efficient. It is designed to be unreliable.

However, there are cases when UDP is an acceptable alternative in an Industrial Internet scenario, for example when receiving streams of sensor data in the proximity network, from a non-critical gauge and the data is still within the acceptable threshold. The controller receiving this stream of data will drop it until it detects a change that crosses a predetermined threshold that will trigger an action.

In this case, UDP is an appropriate low overhead connectionless protocol better suited to non-critical data exchange.

Reliable Transport Protocol (RTP)

This is the favored protocol in VoIP and video streaming applications as it mitigates many of TCP/IP's failings when handling real-time IP traffic in the context of IT systems.

CoAP (Constrained Application Protocol)

CoAP was specifically designed for web transfer in constrained networks and its specialization has found many useful roles in M2M and IoT environments running on constrained devices, such as in the smart building use-case. CoAP is an essential protocol in the proximity networks were it caters to the needs of thousands of constrained devices that don't have the capacity or resources to run HTTP for web transport and transactions.

As CoAP's specialization is working on constrained hosts and networks, it as a very small footprint and can operate on devices that have as little as 10KiB of RAM. However, this doesn't really curtail CoAP's functionality and it works in a manner very similar to HTTP. Indeed, they share a common restful model, using simple commands such as GET, PUT, POST, and DELETE. Similarly, a server in a CoAP environment exposes services as a URL. Therefore, the conceptual models of CoAP and HTTP are so similar that it is easy to integrate them via a proxy when transitioning from a constrained network to a high-bandwidth LAN or WAN.

Moreover, just like HTTP, CoAP can carry just about any type of payload. However as the protocol is purposed to be lightweight, this typically prevents verbose languages like XML, preferring the lower overhead of JSON. Despite having so lightweight and small a footprint, CoAP has strong encryption through DTLS, which still runs on even the smallest end-node sensors.

In order to work across a constrained network (low power and low bandwidth), CoAP doesn't use TCP or anything with complex transport protocols. Instead it utilizes UDP. The UDP protocol is stateless, and therefore does not require a session between endpoint and server, which saves resources. However it is also unreliable. CoAP encapsulates the UDP packet with a fixed 4-byte header, which keeps messages small and avoids fragmentation on the link layer. This allows servers to operate in an entirely stateless fashion.

CoAP is not just confined to the proximity network and it is more often used as a full end-to-end solution connecting servers with end-node devices across the access and proximity networks. One very common lightweight combination is the teaming of CoAP with 6LoWPAN, which is a constrained version of IPv6, and the publish/subscribe protocol MQTT, to provide very efficient end-to-end communications over an IPv6 network. This makes it ideal for IIoT use-cases.

CHAPTER 8

Middleware Software Patterns

The Industrial Internet requires real-time detection and reaction if it is going to work in time-critical environments, such as controlling machinery on a production line or monitoring the status of medical equipment connected to patients in a hospital. Therefore, the applications in the operations and management domain must receive notification of any change in status—crossing a predetermined threshold—immediately. This is also desirable in the world of the consumer IoT; however, with so many sensors to monitor, how can we do this?

There are two ways that applications can detect changes in an edge device's status, and the traditional method is to poll devices and request their status or simply read a value from a pin on an integrated circuit. It amounts to the same thing; we are performing a predetermined batch job to poll each device.

The problem with this method is it is complicated to program and to physically connect these entire point-to-point devices to server relationships, as it will soon become very messy. Furthermore, it is up to the programmer or the system designer to set the polling interval. As an example, if we wish to manage a HVAC system in a building, we need to monitor the temperature in each

© Alasdair Gilchrist 2016
A. Gilchrist, *Industry 4.0*, DOI 10.1007/978-1-4842-2047-4_8

location, but how often do we need to poll the sensors, to read the current temperature? Similarly, if we are monitoring a pressure pad in the building's security, how often do we need to poll for a change of status?

Presumably, with temperature we would only need to check every minute or so, as there is probably a reasonable amount of acceptable drift before a threshold is breached. With the security pad, I guess we would want to know pretty much immediately that an event has occurred that has forced a change in status. In both these cases, the designer will have to decide on the correct polling time and configure just about every sensor—and there could be thousands of them, monitoring all sorts of things such as smoke, movement, light levels, etc.—to be polled at a predetermined time. This is not very efficient; in fact, it is fraught with difficulties in maintaining the system, let alone expanding it by adding new sensors and circuits.

The second option is to have a change of status at the sensor/device level trigger an event. In this event-driven model, the application need not concern itself with monitoring the sensors, as it will be told when an event has occurred. Furthermore, it will be told immediately and this provides for real-time detection and reaction.

In the scenarios of the temperature sensor and the pressure pad, the designer in an event-driven model no longer has to decide when to safely poll a sensor. The sensor's management device will tell the operation and management application when an event—a notable change in condition—has occurred.

In the world of the Industrial Internet, event-driven messaging is required as it can facilitate real-time reaction in time-critical use-cases. In addition, event-driven messaging simplifies the communication software and hardware by replacing all those dedicated point-to-point links between applications and sensors with a software message bus by using multicast to replicate and send messages only to registered subscribers.

A message bus is a software concept, although it can also apply to the physical infrastructure if there is a ring or bus configuration. More likely, the physical network will be in a hub/spoke hierarchy, or remote point-to-point links, although that does not affect the concept of the publish/subscribe model, as a broker server is deployed to manage subscriber registration to the publish/subscribe server and handle the distribution of messages between publishers and subscribers regardless of their location.

Therefore, publish/subscribe works regardless of the physical network structure, and as a result publish/subscribe greatly reduces the amount of spaghetti code and network connections, thus making the whole system more manageable. Furthermore, using a message bus model allows for easy upgrading of the system, as adding new sensors/devices becomes a trivial process.

The reason upgrading becomes trivial is due to the nature of the event-driven messaging system. In order for it to be effective all sensors/devices, which are publishers connect via a broker—a server designated to manage the publish/subscribe process and distribute traffic across the message bus, to the registered subscribers. Similarly, all applications, which are subscribers, connect to the message bus.

Publish/Subscribe Pattern

By using a publish/subscribe protocol, applications can individually subscribe to published services that they are interested in, such as the status of the temperature and pressure pad in the earlier smart building scenario. The publish/subscribe protocol can monitor individual or groups of sensors and publish any changes to their status to registered subscribers. In this way applications will learn about a change in the service immediately.

In addition, there is the added benefit that applications, which use a publish/subscribe broker, need only subscribe to services they wish to know about. They can ignore other irrelevant services, which reduces unnecessary traffic and I/O demands on servers.

Indeed, the designer will not even have to worry about programmatically informing applications of new sensors/services as the publish/subscribe broker service will announce or remove new or retired services.

Additionally, and very importantly for many IIoT applications (such as monitoring remote vending machines), the broker does not require constant connectivity as it can cache and deliver updates from a publisher when subscribers comes online. A vending machine—a publisher—in a remote area does not need to have 24/7 connectivity to the broker; it can perhaps connect once a day to upload reports on its local inventory, cash box holdings, and other maintenance data. On the other hand if the vending machine cash box reports it is full and can no longer accept more cash transactions, the publish/subscribe protocol can alert the broker immediately. Similarly, the subscriber service need not be online, as the broker will store and forward messages from publishers to subscribers when possible.

Publish and subscribe has several distribution models, but what is common is the de-coupling of the sensor/service from the application and this is a very powerful concept in software called *late-binding*. Late-binding allows new services to be added without having to reconfigure the network or reprogram devices or applications, which makes the system very agile. It is also possible using the publish/subscribe protocols to reuse services, in such a way as to create new or add value to existing services. For example, a new service could be produced by averaging the output of a group of three other services, and this could then be published to any interested subscriber.

However, a publish/subscribe protocol may seem a perfect solution to the many use-cases in the IIoT, but it does have its own constraints, which we need to be aware. For example, in the vending machine scenario, it was not critical that we knew immediately, in real time, that a particular vending machine ran out of a brand of fizzy drink. It may well be annoying but the world is not going to end. However, what if the publish/subscribe model is tracking stock market values?

Now we have a problem. A vending machine out of stock, or with a full cash box, can no longer trade and is going to lose money if it does not alert the operational and management domain in a timely manner. However, if we consider fast changing criteria such as stock valuations, we need to know in real time—a second's delay could result in disaster.

And here is the problem with the centralized publish/subscribe model. By placing a broker server inline between publishers and subscribers, it adds latency. After all the broker service has to handle messages received from publishers and then determine which subscribers require the message. The broker will look up and replicate the message for each subscriber, but that takes time—albeit microseconds—but in the case of the stock market that is not acceptable.

Additionally, the broker needs to perform these functions serially so some subscribers may get updated messages of a shift in stock price before others. Therefore, we need to find a way to distribute subscriber messages in real time. Consequently, there are other methods of deploying the publish/subscribe model that do not require a broker service. One simple method is to broadcast to all hosts on the network the published message using UDP, but that is wasteful as most hosts will simply drop the packets and it is unreliable. The problem here is that it's better to have a message arrive milliseconds late than never receive the message. After all, remember UDP is fire and forget, although it is close to real time so very tempting to use in real-time applications. Therefore, again we need to consider how we match protocols with applications, as they can require different levels of service.

Another benefit of deploying a publish/subscribe protocol is that it not only connects all applications (subscribers) to all sensors (publishers) via the "bus," but it can in some cases importantly connect all devices to all devices, which are connected to the bus. What this means in the context of the IIoT is that devices can interconnect and communicate with one another over the publish/subscribe protocol. This means that devices can potentially cooperate and share resources. For example if every device must have long distant communication technology—for example a 3G modem and SIM—they would be bigger, heavier, and more expensive. If, however, smaller, cheaper, and dumber devices could collaborate with their heavy-duty neighbors, they could share resources allowing the smaller device to communicate through the other's communication channels.

Publish and subscribe and event-driven messenger services greatly enhance the efficiency of IIoT systems and they can be deployed using a number of publish/subscribe models, dependent on application. As with all IIoT protocols, there appears to be many to choose from, so we will discuss each of the publish/subscribe protocols in common use today, including their individual pros and cons and how they compare to each other in the context of the IIoT. The most commonly deployed are:

- MQTT
- XMPP
- AMQP
- DDS

All of these protocols claim to be genuine publish/subscribe protocols that operate at real time and can handle tens of thousands of devices. However, they are very different and some work at different levels. For example, consider that we can generally categorize these protocols as working at the level of device-to-device, device-to-server, and server-to-server. Then we should have a better understanding and expectation of the performance, measured in time, for each protocol.

Figure 8-1 illustrates the performance expectations of each category.

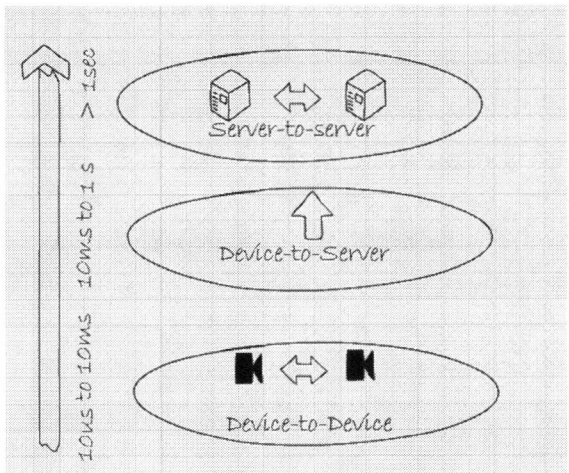

Figure 8-1. Real-time performance expectations

MQTT

MQTT, the Message Queue Telemetry Transport, is a publish/subscribe protocol that's focused on device data collection. MQTT's main purpose is telemetry, or remote monitoring, therefore it's designed to interconnect and retrieve data from thousands of edge devices and transport the aggregated traffic back to the operational and management domain. Therefore, in the general classification, we would consider MQTT to be a device-to-server protocol.

> "Message Queue Telemetry Transport, is an open message protocol designed for M2M communications that facilitated the transfer of telemetry-style data in the form of messages from pervasive devices, along high latency or constrained networks, to a server or small message broker."

Therefore, MQTT is based on a hub-to-spoke topology, although not necessarily so, as it is designed to collect data from edge transducers and send the data collected back to a collection server in the operations and management domain. Because of this design, MQTT does not facilitate device-to-device connections and works in a point-to-point relationship between devices and the collection server. As this is a clear design objective of MQTT, it has few configuration options and does not really require any. MQTT's job specification is simply to collect data from devices and transport the data reliably back to the collection server.

However, that reliable transport requires a reliable transport protocol, so MQTT works over TCP/IP. The implications of that are a full TCP/IP connection is required between the device and the collection server. This is very problematic for a number of reasons; first, it requires the device to be able to support a full TCP/IP stack; second, it requires a reasonable network connection, as the TCP/IP connection must be maintained at all times. Third, using TCP/IP may provide the reliability that MQTT requires but it will affect performance and packet overhead unnecessarily in situations such as a LAN and other non-WAN infrastructures, where there already is a reliable network.

As a result of these design specifications, MQTT is suited to external and remote device monitoring such as monitoring the condition of an oil or gas pipeline, or similar applications where thousands of non-constrained TCP/IP capable sensors/devices require data to be sent to a common server/application. In applications such as remote telemetry monitoring in harsh environments, the TCP/IP protocols reliability is a boon, and the performance is secondary. Consequently, MQTT is not designed for high performance with expected device-to-server figures being counted in seconds.

XMPP (Jabber)

XMPP stands for the Extensible Messaging and Presence Protocol, which is an open technology for real-time communication developed to deliver a wide range of applications. XMPP was developed to focus on the advent of technologies such as instant messaging, presence, and video conferencing, which were dominated by Skype and WhatsApp. There was also the possibility of even entering into collaboration, for developing lightweight middleware, and content syndication.

However, to understand XMPP, we need to understand why it was originally designed. That was for use in instant messaging, to detect presence, and allow people to connect to other people and exchange text messages. Consequently, XMPP was designed to use protocols that make human style communication easier. For example, it uses XML as its native type and an addressing scheme that is intuitive to humans. The addressing format is name@domain.com, which facilitates people-to-people communications, as it is an easily recognizable format common to e-mail and IM programs.

In the context of the IIoT, XMPP may have some useful features such as its user-friendly addressing of devices. It will be easy for a human controller to identify and address devices using a smartphone and a simple URL.

However, we have to recognize that XMPP was initially designed for human usage therefore; it was not designed to be fast. Indeed most deployments of XMPP use polling, or even only-on-demand to check for presence. Therefore, XMPP's performance is based on human perception of real time, which is in seconds rather than micro-seconds.

In addition, XMPP communicates over HTTP riding on TCP/IP, which combined with the XML payload, will make this protocol suitable for smartphones and intelligent devices where the protocol overheads are not a problem. Small low-power devices will not have the processing ability to handle a cocktail of TCP/IP, HTTP, and XML. Consequently, XMPP is best suited for industrial processes that have human-managed interfaces that favor security, addressability, and scalability over real-time performance.

AMQP

The Advanced Message Queuing protocol (AMQP) is not strictly a publish/subscribe protocol but rather as its name suggests a queuing protocol. AMQP comes not from IoT but has its roots in the financial and banking industry. As a result, AMQP can be considered a battle hardened queuing protocol that delivers high levels of reliability even when managing queues of thousands of messages, such as banking transactions.

Due to its roots in the banking industry, AMQP is designed to be highly reliable and capable of tracking every message or transaction that it receives. As a result, AMQP operates over TCP/IP but also requires strict acknowledgment of message receipts from the recipient.

However, despite not being a true publish/subscribe protocol, AMQP is highly complementary to other publish/subscribe protocols, as it delivers highly reliable message queuing and tracking, which is a requirement in some IIoT use-cases. AMQP is also deployed at the server level in the operations and management domain to help with analytics and data management.

DDS

The Data Distribution Service (DDS) targets, in contrast to the other publish/subscribe models, devices that directly use device data. For example rather than being a device-to-server protocol, where servers harvests data from devices in a point-to-point or star topology, as with MTTQ, DDS distributes data to other devices on a bus so it is considered a device-to-device protocol.

While interfacing with other devices is the design purpose of DDS and this means fast communication and collaboration between devices on the same segment, the protocol also supports device-to-server interaction.

DDS's main purpose is to connect devices to other devices. It is a data-centric middleware standard with roots in high-performance technologies, space exploration, defense, and industrial embedded applications. DDS can be remotely accessible and efficiently publish millions of messages per second to many simultaneous subscribers. Furthermore, DDS can store and forward messages, if a subscriber happens to be offline.

The concepts behind how DDS handles the publish/subscribe model is that devices require data in real time as devices are fast. In this context, "real time" is often measured in microseconds. This is often because in IIoT scenarios devices will need to communicate with many other devices in complex ways, to fulfill time-critical applications. Therefore, TCP/IP's slow and reliable point-to-point connection-orientated data streams are far too slow and restrictive. Instead, DDS offers detailed quality-of-service (QoS) control, multicast, configurable reliability, and pervasive redundancy. DDS also provides ways to filter and publish to specific subscribers and they can be thousands of simultaneous destinations, without the time delay of the broker model. DDS can also support lightweight versions of DDS that run in constrained environments such as on low-power devices.

DDS is capable of meeting the demands of high-performance IIoT systems because DDS implements a direct device-to-device "bus" communication with a relational data model. This relational data is called a "data bus" and it is similar to the way a database controls access to stored data. In so much as it efficiently controls access to and ensures the integrity of the shared data, even though the data is accessible by many simultaneous users. This exact level of data control in a multi-device network is what many high-performance devices need to collaborate as a single system.

DDS is designed for high-performance systems so it's a perfect match for IIoT applications such as manufacturing systems, wind farms, hospital integration, medical imaging, asset-tracking systems, and automotive test and safety.

Other publish/subscribe protocols use TCP/IP for reliability. But TCP/IP needs, in order to provide reliable communication, that end-to-end session to be established in order to transmit data at that time. UDP, on the other hand, is connectionless. It fires and forgets. UDP is quick but is inherently unreliable.

However, what if you do need to have reliable data transfer but a connection is just not feasible all of the time? UDP might seem a good prospect, as it is fast and UDP can broadcast to many receivers without too much overhead. However, you don't want to lose valuable data when communications are not possible. Remember that UDP sends once and that is it. Similarly, under conditions where establishing a connection is not possible, TCP/IP will be unable to transmit data. It is only after it establishes a session will it transmit current data. Therefore, the ideal protocol for IIoT scenarios would be one that has store and forward capabilities.

Delay Tolerant Networks (DTN)

This was the problem that NASA encountered when sending space probes into space. TCP/IP was not acceptable as there would not always be a means to form a connection with the space probe, and UDP was also useless as valuable data would just be lost, on a fire and forget basis. The problem that NASA had was how they could ensure that even though there was no radio connection with the probe—for instance, when it disappeared behind a planet—that the data generated by the probe, which they could store locally, could be delivered to ground control in its entirety, once it came into radio contact.

The result of their research was the adoption of the protocol, DTN, which has among its other attributes, store and forward capabilities. DTN is the preferred solution for very remote and mobile IIoT devices and applications and it forms the basis of the middleware protocols that we use today in real-time Industrial Internet situations.

Chapter 8 | Middleware Software Patterns

DTN turns out to be perfect for many IoT applications as it works on a store, carry, and forward basis (see Figure 8-2). For example, the sending node is the sensor in IoT, and it collects and stores the data (store) until it comes into contact (carries) with another node, to which it can pass on the information (forward). The recipient node then stores the data and carries it on its travels until it can forward the data to another node. In this way the data gets passed up through the network until it reaches its destination. The point being here that large variable delays are tolerated as the primary concern is delivering the data to the destination over an intermittent and unreliable network.

Figure 8-2. DTN works on a store, carry, and forward basis

DTN has many applications in the IoT world, for example interplanetary Internet, wildlife monitoring, battlefield communications, and Internet communication in rural areas. When rural villages have no communication infrastructure, communication can still be possible albeit slowly. How it works is the village builds a booth that contains a PC and communication equipment running DTN protocol. The villagers' messages are stored locally on the DTN device. The local bus that services the region and interconnects the rural villages carries Wi-Fi routers that connect/peer to the DTN equipment when they come into range.

As a result the messages are forwarded from the DTN village terminal to the data equipment on the bus. The bus then carries the data with it on its journey from village to village, collecting and storing more data as it goes. Eventually when the bus returns to the town or city depot, it forwards the data it has collected on its journey to an Internet connected router and the data is delivered to the eventual destination over the Internet. The return Internet traffic is handled in a similar manner, with the bus delivering the return traffic (messages) back to each village on it next journey. This is simply a digital version of how mail was delivered once upon a time.

DTN can store-carry-forward data across a long distance without any direct connection between source and destination. Although it could be said that in the previous example a bus was the path between the source and the destination and wasn't random. When handling communication in random networks, for example when performing wildlife tracking, there is no regular bus passing by so we have to make use of other relay nodes to store, carry, and forward the data. When we use these mobile ad-hoc networks, there are several tactics that can be deployed to increase efficiency. One method is called the *epidemic technique,* and this is where the holder of the data passes it on to any other relay node that it comes across. This method can be effective but wasteful of resources and high on overhead as there could be multiple copies of the data being carried and forwarded throughout the network.

One solution to the epidemic techniques inefficiency is called Prophet, which stands for Probabilistic Routing Protocol using History of Encounters and Transitivity. Prophet mitigates some of epidemic's inefficiency by using an algorithm to try to exploit the non-randomness of travelling node encounters. It does this by maintaining a set of probabilities that the node it encounters has a higher likelihood of being able to deliver the data message than itself.

Spray and Wait is another delivery method that sets a strict limit to the number of replications allowed on the ad-hoc network. By doing so, it increases the probability of successfully delivering the message while restricting the waste of resources. It does this because when the message is originally created on the source node, it is assigned a fixed number of allowed replications that may co-exist in the network. The source node is then permitted to forward the message only to that exact number of relay nodes. The relay nodes once they accept the message then go into a wait phase whereby they carry the message—they will not forward it to other relays—until they directly encounter the intended destination. Only then will they come out of the wait state and forward the message to the intended recipient.

As we have just seen with DTN, IoT can work even in the most remote areas and inhospitable environments. Because it does not need constant connectivity, it can handle intermittent ad-hoc network connections and long delays in delivery. However, that is in remote areas transmitting small amounts of data within specialist applications—wildlife tracking and battlefield conditions, for example. In most IoT applications, we have another issue, and that is handling lots of data—extremely large amounts of data.

Therefore, for most Industrial Internet use-cases, the emphasis will be on reliable scalable systems using deterministic, reliable protocols and architectures such as the ones currently deployed in M2M environments in manufacturing settings.

CHAPTER 9

Software Design Concepts

The Industrial Internet will require the convergence of mobile and social devices, the cloud, and Big Data analytics to be successful. The traditional M2M architecture used was based on the principles of the SOA, and now with the introduction of the IIoT, SOA is more important than ever for delivering services, providing insight, and integrating systems. By applying service-oriented architecture principles to IIoT architecture, a business can manage and govern business and IT transformation. The benefits range from the seamless integration of machines, devices, and services, as well as cloud-enabled software, infrastructure, and platforms for services and solutions, which provides for holistic business insight. SOA also provided the agility to externalize APIs. SOA integrates the domains of the IIC reference architecture, such as the control (OT), operations, and enterprise (IT) domains, among others, with the Internet of Things.

In the context of the IIoT, SOA is "simply good design". It rests on a solid foundation of technology and practices that have supported industrial M2M architectures for a long time and are therefore well understood and trusted in industrial businesses and factories.

API (Application Programming Interface)

An API is a programmable interface to an application. The types of API we see regularly today are those that provide interfaces through web services such as SOAP and REST. These are the service-oriented and web-based APIs that have become the popularly acknowledged norm nowadays for interfacing with web-based and mobile applications.

In SOA (service-oriented applications) environments, where we are able to communicate with higher applications such as CRM and ERP, an API is a rather a complicated concept. These programmable interfaces were complex and issued by the software vendor, typically with limited functionality. This was more to do with the vendor retaining control of development and being able to up-sell functionality and integration with other enterprise applications and databases as an additional service. However, as IT shifted away from SOA to web and mobile applications, APIs have also become a relatively simple template that we use to communicate between our mobile and web applications and with other backend databases and enterprise applications. Let's break it down by looking at each of its parts.

Let's first look at an API's component parts.

Application

If you have a laptop, tablet, or smartphone, you are well acquainted with what applications are, i.e., the tools, games, social networks, and other software that we use every day.

Programming

This is the coding that software engineers use to create all the software that make up our applications on our devices.

Interface

An interface is a common method for users, applications, or devices to interact. If we share this interface between two applications or programs, that interface provides a common means for both to communicate with one another.

Therefore, an API is essentially a way for programmers to communicate with a certain application through a software-defined template.

API: A Technical Perspective

For an API to work it must be configured and detailed by the application programmer and typically the database designer. This is because normally you are injecting data from one application into the host application, which means it needs to be stored permanently in the database. Therefore, an API can best be described as being:

> "…a precise specification written by providers of a service that programmers must follow when using that service The API describes what functionality is available, how it must be used, and what formats it will accept as input or return as output."
>
> —Kevin Stanton (API Chapter Manager at Sprout Social)

As a result we can consider an API as being a specified template for inputting or retrieving data into an application or its database.

API Analogy

In order to have a clearer understanding of how the API (template) is used, let us look at an appropriate analogy. Consider for the sake of argument that every time you want to submit or access a set of data from an application, you have to call the API. However, there are certain rules that you must follow and the programmers determine these rules when they construct the format of the API template. After all, when designing the API the programmer will have to determine that there is only certain types of data the application will let you access. Similarly there is only specific data it will allow you to input to the database, so you have to communicate in a very specific protocol and format—a language unique to each application.

To help understand this concept, imagine an API as the security guard between a programmer and an application. The security guard accepts the programmer's requests and checks for authentication. The security guard also supplies the programmer with a form with the correct format for the request and the data. When the programmer completes the form with their request and the details of the request criteria, i.e., what data they wish to store or retrieve, the security guard will check the format is correct. If the presentation is correct, the security guard will forward the request to the application. The application subsequently performs further authentication and authorization control and if the request passes its security checks then the application will allow the request/input, and the application returns or stores the data too or on behalf of the programmer.

Example of an API

A company may give its staff the ability to run reports on their own data, reports, or social media from the company databases. However, in order to run those database queries, a SQL query must be entered against the database.

It is not feasible for the users to send an e-mail to the Database Administrators (DBA) whenever they need that information, and for security and operational reasons, the DBAs cannot give users the access to run the query themselves.

This is where the API comes into play. The database administrators have written unique SQL queries in scripts that trigger whenever a user makes an API call. The database can then return requests for data from the store to the user in real time. Consequently, whenever people request any data analytics using the API, the results are returned without the users having to know any coding or SQL or even interface directly with the database system.

What Is an API Call?

Whenever a user, developer, or system requests information from an application or wants to post information to another application, they need to make a request to that API, using the preformatted template. Of course, having external users and systems make requests to the API requires security measures to be in place to limit bandwidth, usage, as well as illegal requests. Web-based API are URLs, which are used to activate the call using common HTML commands, such as PUSH and GET. These URLs are not activated by clicking on them as they will typically return just a blank screen (unless they have been constructed to send data back to the web page) so they typically are used within programs and scripts. As a result, programmers can build APIs using a whole array of languages, PHP, Ruby, Python, etc. The programmer can then imbed and call the API URL from within the programming language scripts.

Why Are APIs Important for Business?

Let's discuss next why APIs are so important for businesses.

Businesses Create Apps with APIs

APIs are the way business can communicate real-time with applications. Accountants can run reports in the morning for the previous day by a touch of a key and sales can post figures at the end of day in a similar fashion. APIs are the way we can directly communicate with applications without knowing how to write queries or code. Developers create APIs that we use as templates to upload sales figures from a spreadsheet or from a formatted file.

Industry 4.0

Business People Use External APIs

APIs allow businesses to interact and collaborate through social media and external applications such as Twitter, Facebook, and Amazon. APIs also allow users to collaborate through Dropbox, Box, and Google Apps, among many others. Google and other web-scale giants also make their APIs open and available for the public use. For example, a programmer can use Google Maps and Geo-locator API and then use these in their own applications. Developers can make whole mobile apps simply by linking open APIs from Google or Yahoo by using some glue code; these are called "mash-ups".

Businesses Rely on Open APIs

Open APIs are hugely important for providing access to third parties and trusted partners by giving them secure access to data and for project collaboration. Service-oriented architecture and enterprise-scale applications made open APIs available to developers, partners, and IT for system integration, innovation, and development purposes. Open web-based APIs are open to everyone.

Web Services

There are many types of APIs and they can be constructed in many languages but the ones typically used today in service-oriented architectures and modern web and mobile-based applications are web service APIs, such as SOAP and REST.

SOAP (Simple Open Architecture Protocol) is a standards-based web services protocol originally developed by Microsoft that has established itself in the SOA environment as the web service of choice. However, the name SOAP is a bit of a misnomer as SOAP really is not as simple as the acronym would suggest. It really is because of its established use in SOA and the fact that it has been around for a long time that developers and programmers are experienced and comfortable with its use. SOAP therefore enjoys all of the benefits of its longevity.

On the other hand, REST is relatively new and is designed to be simple or at least simpler than SOAP. REST was designed for web and mobile applications and to fix some of the inherent problems that exist with SOAP when used in a web environment. The REST design criteria are to provide a truly simple method of accessing web services through a URL and without any of the XML baggage of SOAP. However, sometimes SOAP is actually easier to use.

The reason for that is SOAP web-service APIs are well documented through the Web Services Description Language (WSDL). This is another file associated with SOAP, and it provides a definition of how the web service works, through the XML template for the API. The WSDL details in readable form the format, including the specifications, requirements, and options used by the service. Consequently, the reuse of SOAP web services is common, but it might not be efficient, in so much as you may retrieve too much data—but it is trivial. By having WSDL provide a definition of how the web service works when you create a reference to it, the IDE can completely automate the process. Therefore, the difficulty of using SOAP depends to a large degree on the language you use.

On the other hand, sometimes REST is not so easy to understand as there is no corresponding documentation like WSDL, the features are determined within the code itself, and this has problems of its own, for non-programmers.

However, SOAP and REST do share similarities as they both can work over the HTTP protocol. SOAP can work over other network protocols, even SMTP. SOAP because of its standardization and its ubiquitous use in the SOA environment is a more rigid set of messaging patterns than REST. The rules in SOAP are important because without these rules, it would be unable to achieve any level of standardization. REST as an architecture style does not require processing and is naturally more flexible.

Both SOAP and REST rely on their own set of well-established rules that everyone has agreed to abide by in the interest of exchanging information. Therefore, both techniques have issues to consider when deciding which protocol to use.

A Quick Overview of SOAP

Microsoft originally developed SOAP to replace older technologies that do not work well on the Internet such as the Distributed Component Object Model (DCOM) and Common Object Request Broker Architecture (CORBA). These technologies were deemed unsuited to the Internet because they rely on binary messaging. However, SOAP relies exclusively on XML to provide messaging services. Consequently, the XML messaging that SOAP employs works better over the Internet. An interesting feature of SOAP is that it does not necessarily have to use the Hypertext Transfer Protocol (HTTP) transport, as it can run over other network protocols.

After an initial release, Microsoft submitted SOAP to the Internet Engineering Task Force (IETF), where it was standardized. From the outset, SOAP was designed to support and accommodate expansion, so it has all sorts of support modules and options associated with it. Because of having so many optional features, SOAP is highly extensible, but you only use the pieces you need for a particular task. For example, when using a public web service that is open and freely available to the public, you really do not have much need for WS-SECURITY.

One of the issues with SOAP is that the XML, which it uses to make requests and receive responses, can become extremely complex. The issue here is the fact that in some programming languages, the programmer will be required to build those XML defined requests manually. Additionally, the issue of manually programming the requests becomes problematic because SOAP is intolerant of errors. Therefore, some Java developers in particular found SOAP hard to use, as working with SOAP in JavaScript is cumbersome because you must create the required XML structure absolutely every time. This which means writing a lot of code to perform even the most trivial tasks. However, the issue with JavaScript is not always the case and some languages make light work of handling SOAP and XML. SOAP can provide shortcuts to the language, which can help reduce the effort required to create the request and to parse the response. In fact, when working with Microsoft's own .NET languages, you never even see the XML.

SOAP may be intolerant of errors, but ironically one of the most important SOAP features is built-in error handling. If SOAP discovers a problem with your request, the response contains error information that you can use to fix the problem. This feature is extremely important as in many cases you may not own the web service, in which case there would be no indication as to why things did not work. The error reporting even provides standardized codes so that it is possible to automate some error-handling tasks in the code.

A Quick Overview of REST

In comparison to SOAP, REST provides a lightweight alternative for a web service API. How REST differs is that instead of using XML to make a request, REST relies on a simple URL using basic HTTP commands. In some advanced or complex situations, REST may have to provide additional information, but most web services using REST rely exclusively on obtaining the needed information from the URL. Because, REST uses an URL approach to call the API, it can use four different HTTP 1.1 verbs (GET, POST, PUT, and DELETE) to perform tasks.

One major benefit for developers and programmers alike it that unlike SOAP, REST does not have to use XML to provide the response. REST-based web services can output the response back to the program as data in Command Separated Value (CSV), JavaScript Object Notation (JSON), and Really Simple Syndication (RSS). Therefore, programmers can obtain the output that they need in a form that is easy to parse within the language they are using to write the application.

Soap versus Rest

SOAP is definitely the heavyweight choice for web service access. It provides the following advantages when compared to REST:

- Language, platform, and transport independent (REST requires use of HTTP)
- Works well in distributed enterprise environments (REST assumes direct point-to-point communication)
- Standardized
- Provides significant pre-build extensibility in the form of the WS* standards
- Built-in error handling
- Automation when used with certain language products

REST is easier to use for the most part and is more flexible. It has the following advantages when compared to SOAP:

- No expensive tools require to interact with the web service
- Smaller learning curve
- Efficient (SOAP uses XML for all messages; REST can use smaller message formats)
- Fast (no extensive processing required)
- Closer to other web technologies in design philosophy

Caching

One web service API behavior that should be noted when using REST over HTTP is that it will utilize the features available in HTTP such as caching, and security in terms of TLS and authentication. However, this might not be a good thing. Depending on the application, it may be a very bad option. For example, designers should know that dynamic resources should not be cached. This is because the resources are changing in real time, so caching the resource is a really bad idea. For example, there is a REST web service that will be used to poll stock quotes when triggered by a stock ticker. It is important to understand that stock quotes are highly likely to change per milliseconds, so if a request for a stock price for BARC (Barclays Bank) is polled and published there is a high chance that on a subsequent poll the quote received will be different.

However, if REST is using the HTTP cache feature, it will return the same stock value that was polled initially and stored in cache, again in subsequent polls. This shows that we cannot always use the caching features implemented in the protocol when working with dynamic content and resources. HTTP caching can be useful in client REST requests of static content but the caching feature of HTTP is not suitable for dynamic requirements; SOAP API is the better choice.

HTTP Verb Binding

HTTP verb binding is another feature worth discussing when comparing REST and SOAP. Much of public facing API referred to as RESTful are more REST-like and do not implement all HTTP verbs in the manner they are supposed to. For example, when creating new resources, most developers use POST instead of PUT. Even deleting existing resources are sent through a POST request instead of the DELETE command.

SOAP also defines a binding to the HTTP protocol. When binding to HTTP, all SOAP requests are sent through a POST request.

Security

Security is rarely mentioned when discussing the benefits of REST over SOAP. The reason for this is that REST security is provided on the HTTP protocol layer such as basic authentication and communication encryption through TLS. SOAP security is well standardized through WS-SECURITY. HTTP as a protocol is not secured, therefore web services relying on the protocol need to implement their own rigorous security. Security goes beyond simple authentication and confidentiality, and also includes authorization and integrity. When it comes to ease of implementation, SOAP is the web service at the forefront.

Microservices

The use of microservices and web APIs has become very popular in web and cloud-based applications and they are both ideally suited for the IIoT.

The term *microservice* has no standard, formal definition; however, there are certain characteristics that identify them. Essentially, there is a distinguishable microservice architecture, which provides a method of developing software applications as a suite of small, modular services. These independent services run a unique process and communicate through a well-defined, lightweight web service or other mechanism in order to deliver a specific result.

What makes microservices valuable are that applications can be constructed that decouple the underlying complexities of the host systems from the application's purpose. If we take a look at a client-server web application, these are built as monolithic applications, where the server portion of the code handles HTTP requests, executes logic, and retrieves or posts data from/to the database. The problem with this style is that any changes will require a new version of the entire application, as each function is intertwined and not readily isolated and updatable. With microservices, we see the introduction of a different approach, one that is ideal for the IoT, and by extension the Industrial Internet.

The microservices approach is to build your own or consume an open microservice, which is specific to a function. Programmers then construct applications using these independent modules called microservices, which the programmer can scale and modify in isolation, leaving the rest of the application untouched. By not compromising the overall integrity of the application, programmers can tweak an individual service before redeploying it. Similarly, they can add or customize a service for a new device type of model without affecting any other service in the application, for example, any existing device type's user experience.

Microservices are scalable, adaptable, and modular, making them ideal for cloud-based applications that have to provide dynamic, yet consistent user experience across a wide array, and ever changing, list of devices.

To clarify the benefits of microservices, and their scalability we can look to many large web-scaled companies, such as Netflix, Amazon, and eBay. All of these web giants have migrated over the last decade to microservice-based applications. Netflix, for example, receives around one billion API calls per day from over 800 different types and models of devices. Amazon similarly receives countless API calls per day from an array of applications from a multitude of device types. These companies could not possibly have supported and scales to this capacity using the monolithic or two-tiered architectures; it was only possible using a microservices-based architecture.

As microservice architecture removes the constraints of having the entire application hosted on one server, this enables developers to allocate resources more efficiently to those parts of the application that needs them. However, the obvious penalty is that by separating the business functions and services, there needs to be a communication system between the services to perform these remote calls and this requires additional integration overhead. Integration is always a tricky business so having more of it to do is not particularly advantageous. However, if you're communicating between services, then either HTTP web services or subscribe/publish messaging protocols can be used to simplify the requirements of security, performance and reliability. On deciding which to use, it is good advice to deploy HTTP web service when asking a question that is requiring a direct answer, and subscribe/publish when making a statement that requires no immediate feedback.

CHAPTER

10

Middleware Industrial Internet of Things Platforms

In the previous chapters, we discussed some of the many diverse communication and wireless protocols, technologies, and software patterns available to designers when building IIoT solutions. As we have seen, there is no one solution for every situation so in most cases we will have to deploy several protocols and technologies to meet requirements and then integrate them in some way. For example, in the proximity network we may find a mixture of legacy and modern technologies; some may for example not support IP or the wireless technologies that we favor in our design so we will have to mix and match technologies. This type of design is referred to as a *heterogeneous design* as it supports several if not many diverse protocols, technologies, and software patterns.

Chapter 10 | Middleware Industrial Internet of Things Platforms

Heterogeneous networks are very common if not the norm. It is rare, other than in Greenfield (brand new) deployments, to have the luxury of a common protocol in what would be described as a homogeneous network. The luxury to be able to deploy endnodes and access networks that can support one protocol such as IPv6 is rare. Unfortunately, heterogeneous networks complicate network designs, as we have to cater to diverse protocol interfacing, transportation, and bi-directional translation of data as they will typically be framed in different formats.

Recall that often the concepts of the IIoT are presented at such a high level that the underlying technologies are almost irrelevant, and it is at this point when we are forced to address how we integrate, administer, and manage all these heterogeneous networks that those underlying complexities suddenly loom large.

Earlier chapters discussed how at a very high-level we build an IIoT network. To recap, in the most simplistic terms, we connect transducers (sensors and actuators) to gateways bordering proximity networks and then transport the data back via an access network for aggregation and translation, before ultimately processing and/or storing the data locally or sending it to the cloud. Typically, the IIoT architecture is described in project meetings or IIoT presentations, and the underlying complexities are abstracted from the business goals. Someone will step up to the whiteboard and draw something like Figure 10-1.

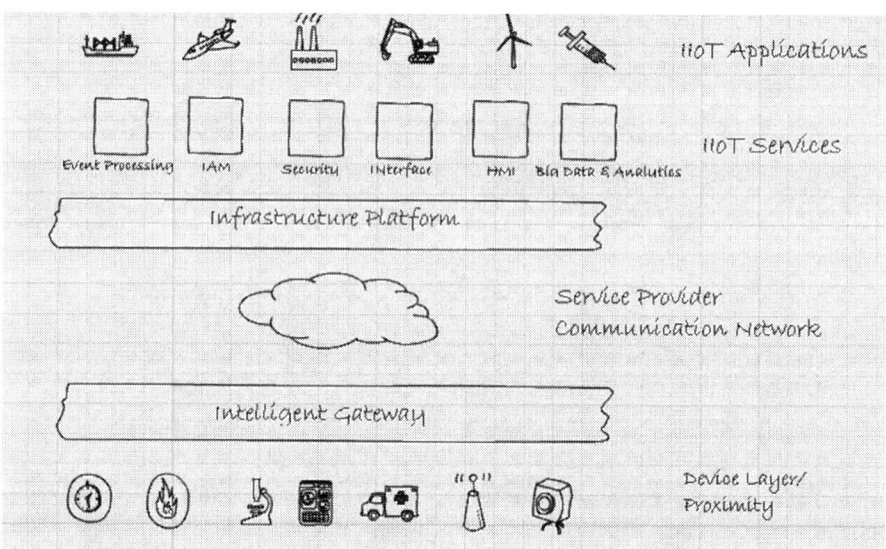

Figure 10-1. IIoT conceptual diagram

Now that is all very well and good; a drawing can make clear how the various components connect and you could feel confident about prototyping or building a proof-of-concept network design. You might even feel confident enough to aggregate several heterogeneous networks. However, one of the key requirements of any good network design, regardless of its technology, is that it should be able to scale. With the IIoT scalability is an issue due to the potential numbers of endnodes. For example, the design criteria may requires thousands if not millions of endnodes (transducers) to be connected. Similarly, these endnodes may be located in thousands of heterogeneous network segments that are not just remotely connected but geographically diverse. The problems that arise are not just how you are going to glue all these heterogeneous components and networks together but how on earth you are going to secure, administer, provision, and upgrade all of these endnodes?

As the mist begins to clear, we begin to have this realization of the underlying complexities of the IIoT. For not only are we going to have to identify and authenticate each and every endnode, which may or may not be contactable (because endnodes may sleep for long periods), we are going to have to apply access control and authorization policies. For example, in a network with thousands of sensors or endnodes, you must be able to identify and authenticate each one. After all, you cannot secure a network if you do not know what you are securing. Similarly, how will we handle additions to the network; for example how do you provision and register new endnodes? Additionally, how can you administer and manage this network? For example, we have insight into the status of endnodes so that we know when one fails or when one requires an upgrade to its software? Finally, how do you detect, upgrade, or provision hundreds of thousands of endnodes in thousands of heterogeneous networks scattered around the country?

Why Do We Need IIoT Middleware?

The high-level diagram drawn on the whiteboard masks the underlying complexity of the technologies that we need to deploy in order to construct an IIoT network. In reality, building an IIoT network infrastructure is an onerous and highly complicated task as we are integrating many different technologies and protocols as well as having to find a way to administer, upgrade, and report on these heterogeneous networks. However, there are solutions that can ease deployment complexity and provide not just the glue to connect all the components, but also a looking glass that will enable us to visualize the network in its entirety, and that is an IIoT middleware platform.

Middleware Architecture

The purpose of IIoT middleware is to provide the integration, connectivity, and data translation that we require between the different technologies and layers. For example, the middleware platform should be capable of handling connectivity from many different protocols and connectors as well as expose and consume API software connectors. Figure 10-2 shows the functionality and interconnectivity of a middleware platform architecture.

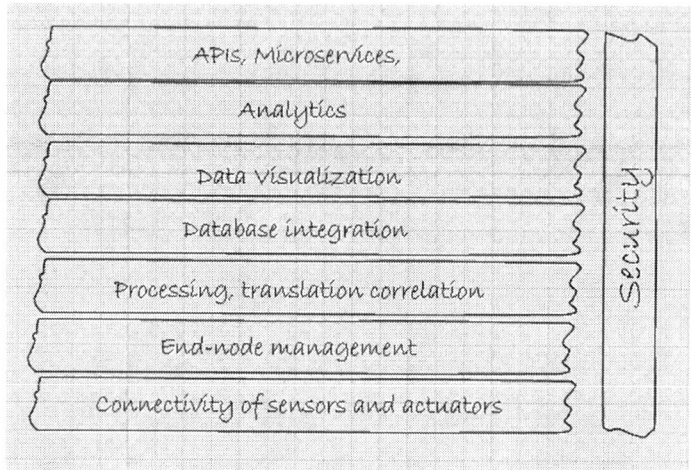

Figure 10-2. Middleware platform architecture

As you can see from Figure 10-2, there are eight components that are desirable in an end-to-end IIoT middleware platform. These elements provide the glue that will allow us to build heterogeneous networks at a large scale.

- Connectivity—This layer provides the means to connect and support the sensors and actuators as they will often have diverse technologies and protocols.

- Endnode Management—This layer provides the ability to identify, authenticate, authorize, and manage all the endnodes in the network.

- Data Processing—This layer provides the data translation, preparation, and correlation of data coming from the sensors.

- Database Integration—This layer provides the connectors between applications and data storage.

- Data Visualization—This layer provides the tools and techniques to visualize the data in a meaningful way, such as through graphs, charts, and events.
- Analytics—This layer provides the real-time processing and analysis of data that is vital for industrial applications as it enables fast feedback from the industrial control systems.
- Front-end—This application level layer provides the connectors, APIs, microservices, and SDK interfaces required by applications and developers.
- Security—The security of the IIoT network is paramount and is addressed at every layer to ensure the confidentiality, integrity, and availability of the network.

However, it is not quite that simple as selecting any IIoT middleware platform as they come in various shapes and forms. Typically, an IIoT middleware platform should bring heterogeneous systems together, provide seamless integration and interoperability while at the same time ensure operational transparency at the application layer. In order to achieve this middleware, whether it be IT or IIoT, you need a mechanism for communication between diverse technologies and protocols, while abstracting the underlying complexities of hardware, software, and communication protocols. The aims are also to aid portability and seamless integration across systems.

Unfortunately, not all so-called IIoT middleware platforms cover all the bases, as some are simply basic connectivity platforms, or IT application mediation platforms for integrating applications with databases. Some IT middleware platforms are actually only communication mediation systems that handle protocol translation and present physical interfaces for physical connectivity. Others are data management middleware that mediate between applications and database back-end systems. Finally, a new breed of IaaS platforms have come about that optimize mobile apps and IoT that could also be described as IoT middleware.

What Should We Look for in an IIoT Middleware Platform?

Thankfully, as with most things regarding the IIoT, these problems are not something new that we haven't encountered before. In fact, not even the scale of the IIoT is a problem.

Mobile phone operators have had this problem for decades and the solution is to adopt a middleware platform that will handle identification, authentication, authorization, and the capabilities to manage the automated provisioning of tens of millions of remote devices. Similarly and perhaps more specifically suited to IIoT deployments, enterprises have adapted these high-end mobile operator platforms to enable BYOD (bring your own device), which allowed them to authenticate and control access to their networks as well as manage tens of thousands of mobile devices, which were not typically under their direct control.

The solution once again was to use a middleware platform that automated, identification, access control, security, provisioning, and reporting amongst other vendor specific functionalities. These Mobile Device Management (MDM) platforms are very similar to the new breed of IoT middleware platforms aimed at SME deployments, and the next section looks at the functionality they provide and how we can deploy them within an IIoT network design.

What Do IIoT Middleware Platforms Do?

The problem is that perhaps hundreds of thousands of transducers will be connected to the Internet, which we will expect to securely replace traditional data sources, and these will become the main source of data for an industrial business. This data could flow from the sensors and perhaps even from the products under manufacture, or from feedback from operational processes, even the operating environment. However, data is meaningless without the ability to turn those bits and bytes into knowledge and then understanding. The challenge of transforming data, those raw bits and bytes, into knowledge is a technological one. However, every challenge also represents an opportunity and many vendors are working to produce IioT-capable middleware platforms to provide the glue to connect all these disparate transducers, applications, and databases.

Middleware is critical for building and deploying IoT applications. This is because middleware platforms provide multiservices such as interfacing of diverse protocols and technologies, translation and identification, authentication and authorization, among many other features. An example of a middleware platform is the open source solution Kaa, which can support connectivity, scalability, administration, and data-streaming analysis for potentially millions of endnodes. Furthermore, middleware platforms provide multiple additional services, including determination of software/firmware levels, quarantining, and remote provisioning and upgrades. Finally, yet importantly, middleware plays a major role in correlating and presenting data for processing and storage. This is very important because in today's industrial applications of the IIoT, we can be dealing with vast amounts of data. For example, Boeing jet engines produce 10 terabytes of data during every 30 minutes of flight.

There are many ways to extract value from the data that results in significant improvements in operational efficiency within industries such as industrial manufacturing, aviation, smart utilities, healthcare, and transportation. However, there are still problems that arise from the rapid growth in the amount of data generated and the business demands for process agility and execution. So what are the constraints that the IIoT middleware need to overcome?

Event processing is a major element in an Internet of Things platform. This is because major amounts of data can stream from sensors, so it's important to derive an understanding of which data is important and which is not. After all, there is little point sending the same data continuously from a restricted device over a limited wireless communication link if the data isn't changing.

However, industrial systems do need to know immediately when an endnode's status changes, with acceptable boundaries, so that they can take immediate and appropriate action. Therefore, industrial and operational technology businesses tend to utilize real-time data streaming and analysis as they require instant feedback in order to stabilize a process that may be wandering out of the acceptable boundaries.

Consequently, an IIoT middleware platform should be designed in order to return the key performance indicators that industrial scenarios demand. For example, when dealing with large number of devices, scalability and reliability is of paramount importance. Middleware provides the enablement platform for building, deploying, and managing scalable IoT applications. Therefore, the fundamental role of middleware in IIoT can be summarized into three key value propositions that a business would require. At a very high level, the requirements are for

- Integration
- Secure, monitor, and report
- Real-time and batch data analysis
- Open source middleware solutions

For many SMEs, middleware can be a very expensive option. Fortunately, there are now open source and commercial middleware platforms that we can deploy to manage these disperse technologies and protocols. Without these open source middleware platforms that integrate and seamlessly translate between diverse technologies and protocols, the Industrial Internet of Things would not be feasible for small medium enterprises that would not be able to afford the large-scale IIoT platforms. For example, for large companies, there are enterprise class solutions available from Siemens, GE, and SAP, but these are typically out of the financial budgets of most SME businesses. So what are the options for SMEs wishing to deploy IIoT?

Chapter 10 | Middleware Industrial Internet of Things Platforms

There are several mature, open source solutions for IIoT middleware:

- Kaa—Provides the tools to build complete Industrial Internet solutions by interfacing transducers, protocols, and applications. Kaa also provides the means to monitor and manage each remote device.

- OpenIoT—An open source middleware platform for implementing and integrating IoT solutions. OpenIoT is designed to be able to connect to and then collect and process data from just about any transducer regardless of its protocol. OpenIoT can then stream the collected data to the cloud and analyze and visualize the collected data.

- Alljoyn—A platform that makes it easy for devices to discover and communicate with one another regardless of the protocol, manufacturer, or the transport layer.

- Mango—One of the most popular IoT platforms due to its traditions in the M2M, industrial control, and SCADA industrial environment, where it has built up a huge following due to it ease of deployment, low power use, and ability to host thousands of devices from one low-cost PC.

Open source, though very popular, is not always seen as the best solution in industrial environments where reliability, availability, and robustness could depend on the quality of the product and the technical support. Consequently, many businesses will invest in commercial products in order to buy peace of mind and the technical support they require should things go wrong. Some of the most popular commercial IoT middleware platforms are:

- ThinkWorx
- Oracle Fusion
- IBM Bluemix

These commercial platforms have more built-in features, such as application enablement and development tools as well as network and device management support, and are typically easier to deploy. But they come at a price. However, the cost of licensing these products can often be cheaper than the cost and time spend deploying, testing, and supporting mash-ups of open source packages.

CHAPTER 11

IIoT WAN Technologies and Protocols

The core difference between M2M and the IIoT architectures is the interaction with the Internet. The Internet connectivity allows M2M systems located in remote locations to communicate beyond their natural boundaries. This facilitates collaboration among devices, customers, partners, and supply chains around the globe.

The Internet has revolutionized data communications. Previously, a mesh of expensive, point-to-point leased lines, frame-relay, or ATM circuits had to be leased from a telecommunications service provider in order to manage remote data communications, regionally or internationally. The Internet has transformed the way data communications are handled globally and now all that is required is a single Internet connection to a local service provider, and the Internet will discover remote entities and set up the necessary connections between these hosts. To provide security and to emulate the lease-line and hub-spoke multi-point circuits such as frame relay, companies can use the Internet's backbone to establish VPNs between sites, making provisioning of intersite connectivity highly efficient, reliable, private, and cheap.

© Alasdair Gilchrist 2016
A. Gilchrist, *Industry 4.0*, DOI 10.1007/978-1-4842-2047-4_11

However, that reliability is dependent on the purpose; for example for short occasional VPN connections from road warriors to their head office, Internet broadband VPNs are certainly reliable enough. The problem comes when we need, as is the case with a many enterprise and IIoT applications, a 24/7, 365 days a year connection, to a cloud provider for example. In this specific case, a broadband wireless Internet connection is unlikely to meet the required levels of service reliability and up time. Therefore, we have to understand that not all Internet connections are of the same quality or reliability, so we need to choose Internet connectivity wisely.

The choice of Internet connectivity for IIoT varies considerably depending on its specific purpose; devices for example in the proximity network have different requirements of performance and service levels, to an Internet connection between a factory's operations and management systems and the cloud. The typical WAN choices available in urban areas differ from those offered by service providers in rural areas and the quality, reliability, and throughput will also likely be significantly less.

Therefore, when considering a suitable WAN technology for Industrial Internet connectivity that is practical and financially feasible for projects that require high performance for interconnectivity between remote stations, headquarters, and cloud or data center interfaces, consider the following;

- Availability
- Bandwidth
- Cost
- Management
- Application
- SLA and reliability
- QoS

So what are the WAN technologies that service providers offer?

In some geographical regions (such as in some areas of Africa and Asia), you might find yourself restricted to legacy communication WAN channels such as the following:

- ISDN—Is a circuit-switched technology and it comes in two versions—basic rate ISDN (B-ISDN), which is two basic telephone lines bundled together providing 128Kbps and was commonly installed in small offices and homes for internet access back in the days of dial-up modems. The larger primary rate (PRI) version of ISDN, came in either T1 (USA) or E1 (Europe and most of the world). These are bundles of standard 56kbps telephone channels that make up 1.5Mbps (T1) and 2.0Mbps (E1) respectively.

PRIs were predominantly used in PSTN telephone trunks but they also were used extensively for backhaul in mobile 2G and GPRS networks.
- Frame relay—This synchronous serial technology is still around in many parts of the globe. Frame relay was provided by Telco service providers that built PVC (Permanent Virtual Circuits) across their infrastructure to build global multi-access networks.
- ATM—Another legacy circuit-switched technology that used the same PVC concepts as frame relay. ATM however used small 53-byte cells, with a 48-byte payload, which made overhead to payload ratio unsuitable for most types of data traffic. ATM's synchronous nature and fixed cell size made it ideal for reliable voice and video applications, as performance was deterministic. ATM is most commonly offered via dedicated circuits or through 2G/GPRS data circuits.

In developed countries, the choice of WAN technologies will vary depending on the customer's requirements. The most common ones are:
- xDSL—This is a very popular fixed-line technology offered by Telco service providers because it can utilize an existing telephone line. By using the local loop copper to connect a local DSM modem/router to a DSLAM in the Telco's premises, DSL services can be provisioned cheaply and efficiently. The alternative is to run over fiber, which was once rare and expensive but is now far more common as business demand for higher and higher data speeds appears insatiable. DSL is suitable for IIoT back-end and cloud interconnection as it is fast and reliable. Additionally, multiple xDSL circuits can be bundled using techniques such as software-defined WAN (SD-WAN) to provide greater reliability and throughput by aggregating available bandwidth. One drawback to xDSL is that the advertised bandwidth is shared among subscribers and service providers oversell link capacity, due to the nature of spiky TCP/IP and Internet browsing habits. Therefore, contention ratios—the number of other customers you are sharing the bandwidth with—can be as high as 50:1 for residential use and 10:1 for business use.
- SDH/Sonnet—This optic ring technology is typically deployed as the service provider's transport core as it is provides high speed, high capacity, and highly reliable and fault-tolerant transport for data over sometimes-vast geographical regions. However, for customers that require high-speed data links over a large geographical region, typically enterprises or large company's fiber optic

rings are high performance, highly reliable, and high cost. Sonnet and SDH are transport protocols that encapsulate payload data within fixed synchronous frames. By aggregating frames, SDH/Sonnet can handle higher payloads, and therefore higher throughput. Ethernet can be carried over SDH/Sonnet due to its protocol neutrality and high-speed variants. 10G, 40G, and 100G map directly to OUT-2, OUT-3, and OUT-4, respectively. SDH/Sonnet is still widely used with Sonnet used in the United States and SDH in Europe and the rest of the world.

- MPLS—IP/MPLS is the modern and most common backbone technology of service providers today. MPLS has replaced the legacy leased-lines, frame-relays, and ATMs in most countries but not all. Therefore, when you buy a private leased line or LAN, you are most likely offered a provider's MPLS private wire or LAN. MPLS uses a fast switching technology that uses labels to encapsulate other protocols. MPLS can encapsulate and switch anything, which makes it hugely flexible in Telco service provider networks. MPLS can switch Ethernet across vast areas using layer-2 VPNs or route traffic using layer-3 VPNs. Ethernet services are sold as virtual leased private lines or LANS. MPLS is fast, extremely reliable, secure, and easy to manage from a user perspective. However, it can be slow to provision, unavailable in some regions, and expensive.

- 3G/4G/LTE—Carrier broadband has come a long way in the last decade with huge improvements in data handling over mobile operator broadband networks, since the days in the early to late 2000s of 2G and GPRS. GPRS does however have IIoT relevance, as it can be useful still in device interconnection in remote areas. However, with 2G networks being decommissioned it might not be around much longer. 3G/4G and LTE have broadband capabilities with high data throughput availability. It is 3G/4G and LTEs wide coverage footprint that makes it so enticing for IIoT urban and rural applications. However, it has its drawbacks when used for device connectivity as it requires high power, so has poor battery life. Those failings can be mitigated by connecting many devices to a local 3G router, for example in a hub-spoke topology, which can make better use of the service available. However, because mobile networks are designed for voice and broadband data the price of a data plan and SIM for a remote router can be prohibitive. Mobile operators are beginning to offer IoT service plans so that might change, but even if the price does fall, the inherent design of the networks will not, and that will still make them technically inefficient for devices that communicate only tiny amounts, intermittently.

- DWDM—The most advanced core transport optical network and typically only affordable to service providers and huge enterprises. DWDM carries data arriving from different sources over the same optical carrier simultaneously. It manages this by multiplexing the data sources using an optical add/drop multiplexor and placing each on a different optical wavelength for transport across the optical fiber. By utilizing optical filters, the wavelengths travelling at different frequencies can be separated at the destination and the data extracted. DWDM greatly enhances the capabilities of optic fiber because, by using different wavelengths to transport different data sources independently but simultaneously across the same strand of optical fiber, they create in effect virtual fibers. DWDM can currently support 80 different wavelength channels, with each channel carries data at bit rates of 2.5Gbps. DWDM is used typically in service provider backbones, mobile core networks, metro-Ethernet networks, and even for connecting enterprise data centers using layer-2 switching over DWDM. DWDM is the premier high-capacity, high-performance transport mechanism today and it is becoming popular for short and medium point-to-point links as well as for long distance core rings.
- FTTX—Fiber to the home/curb/street are all variants of a service provider's fixed-line WAN portfolio for bringing high performance applications, such as IPTV and triple play, over fiber to the premises. Fiber optic cable have almost unlimited capacity and throughput with data travelling at the speed of light through the glass conduit within the fiber cable. FTTX is delivered in a number of ways from the service provider to the building and some will use active optical networks, which provide for a single dedicated fiber to be run. That is likely to be prohibitively expensive though in highly secure environments and applications this might be acceptable. The more often used FTTX configuration is to deliver FTTX via PON (passive optical networks). PON uses passive optical splitters to split the optical signal so that it can be

- shared among customers. The optical splitter can create anywhere from eight to 64 branches, although more are possible.

- Cable Modem—This technology is more suited to a consumer IoT application, but it will have no doubt numerous use-cases in IIoT. Cable technology requires a cable modem that provides high-speed data transmission over coaxial or fiber lines that transmit cable television, or triple-play services.

- Free Space Optics—Sometimes there just is not any available communications infrastructure, fixed or wireless. This is often the case in remote rural areas in developing nations. Although there are national operators of mobile and fixed line services, for financial reasons they don't provide coverage. The only recourse is to build your own network. Free space optics can be the ideal solution for building point-to-point line-of-site high performance Ethernet links. With speeds up to 10Gbps and range up to 3-5km, these are very affordable ways to backhaul traffic or connect remote locations to a nearby network. However, FSO is very vulnerable to weather and atmospheric conditions—fog and sandstorms being FSO killers. As a result, FSO systems tend to be hybrid with built-in broadband radio for fail-over in the case of loss of signal on the FSO link. Although the fail-over works well, the throughput drops to often unacceptable levels of around 10-20 Mbps.

- WiMax—A high-speed RF microwave technology that provides up to 1Gbps and was designed as a last mile broadband access to rival DSL and cable. There was a lot of interest in WiMax as it was considered a rival to the LTE mobile technology; however much of that interest has waned due to LTE's dominance. WiMax works similarly to Wi-Fi but over much greater distances, which makes it suitable for IIoT applications, where service is available. WiMAX was designed for mobile broadband access, IPTV, and triple play delivery and telecoms backhaul, so it is optimized like 3G/4G and LTE for large throughput. With WiMAX, ranges of 50km are achievable but at the expense of bit-rate. However, if high bit-rate is not a priority, as it is not usually with IIoT remote devices, then even greater distances can be achieved.

- VSAT—Very Small Aperture Terminal is a satellite communications technology used to deliver broadband Internet and private network communications. VSAT is ideal for very remote locations where there may well be no mobile network coverage, such as on ships at sea, oil/gas platforms, mining camps, or for communications in remote field operations. VSAT can be used for Internet service, e-mail, web access, and even VoIP and video. In IIoT use-cases, VSAT can play an important role due to its vast coverage footprint.

These are the typical long-distance WAN alternatives that connect businesses to data centers and could operate as they are configured. However, in IIoT designs, especially in urban and rural areas where hundreds or thousands of sensors may be deployed, we require other technologies that can handle low traffic at affordable prices.

IIoT Device Low-Power WAN Optimized Technologies for M2M

On many occasions, we need to connect edge devices to the Internet that are situated outdoors, such as remotely in urban or rural areas and that have no available connectivity close to hand. This is an everyday problem that designers of IIoT systems have to face in industrial and civic engineering applications.

While there are many diverse IIoT applications for low-power WAN communications, there are a few common denominators that the technology must provide. These prerequisites include:

- Low cost—Cost is always important in IIoT design and it has particularly hefty weighting when considering low-power WAN technologies. The fundamental reason is that when considering remote-access applications, designers will anticipate a requirement for 100s if not 1,000s of endnode devices, sensors, or actuators. If we consider a metropolitan IIoT initiative for smart traffic management, millions of endnodes may be deployed over a wide area. Consequently, designers have to consider the capital and operational costs of a low-power WAN technology when embedded in a device.
- Low energy consumption—Another key design constraint is how power hungry the WAN technology is. As with cost, the design will requires 1000s of endnodes and each one will likely require its own power source,

though some will be able to energy harvest, which would be a major plus. The problem with radio wireless communications is that it takes a lot of energy to transmit a signal and most remote endnodes will likely be supported by mini-button batteries, which need not months but years of life expectancy to be viable. Anything less could prove to be a hugely expensive, logistical nightmare. For instance, imagine tracking, monitoring, and changing a million batteries in the end devices of a metropolitan traffic management system.

- Operating range—The operating range of the WAN technology again has cost and efficiency benefits. This phenomenon can be demonstrated by the necessity of an edge device to connect via an access point or a protocol gateway situated between endnodes and the Internet or an IT infrastructure in an all wireless technologies. The fewer protocol gateways and access points required, the more efficient the design and the cheaper the overall system. Similarly, the greater the RF reach of each access point and endnode, the higher the number of endnodes that can be deployed within that access point's sphere of influence. In an urban or suburban area, an access point's strategic position can result in coverage that can cover a wide footprint, which reduces the overall design costs.

- Small form factor—Not all low-power WAN devices require miniaturization, but in some instances, it is a critical factor in the design. By being lightweight and miniature, radio and antennae can be fitted and concealed easily in just about any use-case. However, small form factors can influence the choice or power source.

- Scalability—A network must be able to scale, as more demands on the network will inevitably come over time. Therefore, the access points should be able to handle some additional and reasonable growth without additional spending on access point infrastructure. Similarly, frequency spectrum is an important consideration especially when operating in the unlicensed frequency bands as noise and interference from neighbor's wireless networks operating on the same frequency bandwidth and channels will cause conflict. Sometimes in the worse cases in urban areas, there may not be free unlicensed frequency channels available.

These are just some of the requirements of a low-power WAN network solution. However, there are many other desirable factors that are application specific and so as a result many low-power WAN technologies have sprung up to cater to specific use-cases, resulting in a range of competing technologies and protocols. A closer look at some desirable attribute for lower power WAN networks reveals why there are emerging technologies required to satisfy diverse design criteria. For example, roaming is the ability of a device to move around the network and cross access-point cells seamlessly without interruption to communications. This is not the same as roaming in mobile network operator's terminology, where a device can seamlessly operate in a partner's network, such as a cell phone being used in a car traveling around a country where it enters other MNO territories. In some use-cases, this of course may be required, but in low-power WAN deployments just being able to move around the device's own home network will suffice.

Some other considerations regarding low-power WAN networks are the technologies ability to handle small and infrequent transmissions. Many applications, especially in the IIoT, monitor remote sensors on devices and these may only require small amounts of data infrequently. For example, a GPS tracking application may only send its location data once every hour. Similarly, some devices may sleep or wander out of network coverage for weeks on end before triggering a small data transmission. These types of devices are inherently unsuitable for 3G/4G or LTE data services, as they will never use the available and expensive bandwidth or service. Consequently, it is often desirable that low-power WAN networks have very limited throughput that handle small message sizes with short transmission intervals, as this can conserve battery life.

Bi-directional communications is also a great thing even though some devices will never require receiving information to perform their daily functions; for example, for a temperature or a light level sensor at least there will probably be a means for remote updates, and importantly for device management, which in IIoT networks is a big issue. Additionally, from a security standpoint there is the opportunity for bi-directional communications to perform two-way authentication of both the access-point and the device. This can prevent rogue devices joining the network perhaps for malicious purposes. Security is also an issue with access points, as rogue access points often pop up in mobile networks so it will be no surprise to see them appear in IIoT networks.

Security is also dependent on the layers defined by the low-power WAN technology as some may provide higher layer services, while others may operate at the lowest MAC/PHY layer and leave the upper OSI layers open as a matter of user choice.

The IIoT encompasses such a broad spectrum of industries and use-cases that no one low-power WAN technology will fulfill every use-case requirement. Therefore, several technologies have emerged that offer different choices. Sometimes there are trade-offs between range and throughput, or battery life

Chapter 11 | IIoT WAN Technologies and Protocols

and operating frequency. However, each technology has its own set of attributes so it should be possible to find one that matches your requirements.

There has been a lot of research and development into producing low-power IoT wireless technologies suitable for remote low-capability devices. The results have been a proliferation of new technologies, some already deployed and tested in the field, while others are still in development but expected to be deployed soon. The list that follows is not definitive as new technologies and protocols are constantly evolving. The main technology players in the low-power WAN field are discussed in the following sections.

SigFox

SigFox LP-WAN is an end-to-end system consisting of the presence of a certified modem and ending with a web-based application. Developers must acquire a modem from a certified manufacturer to integrate into their IoT endnode device. Alternatively, there are third-party service providers that make SigFox-compatible access point networks available to handle traffic between the endnodes and SigFox servers. The SigFox servers manage the endnode devices in the network, collect their data traffic, and then make the data and other information available to the users through a web-based API.

To provide security, the SigFox system uses frequency hopping, which mitigates the risks of message interception and channel blocking. Furthermore, SigFox provides anti-replay mechanisms in their servers to avoid replay attacks. The content and format of data sent in the transmission is user-defined so only the user knows how to interpret their device data.

Name of Standard	SigFox
Frequency band	868MHz/902MHz ISM
Channel width	Ultra narrow band
Range	30-50km (rural),
	3-10km (urban),
Endnode transmit power	-20 dBm to 20 dBm
Packet size	12 bytes
Uplink data rate	100Bps to 140 messages/day
Downlink data rate	4 messages of 8 bytes/day
Devices per access point	1M
Topology	Star
Endnode roaming allowed	Yes

LoRaWAN

The LoRaWAN architecture is a "star of stars" structure of endnode devices connecting through gateways in order to connect to network servers. The way LoRaWAN works is that the wireless hop between the endnodes and gateway uses a Semtech proprietary spread spectrum radio scheme, which they call a *chirp*. The endnodes communicate by chirping the gateway when they have data to transmit. The radio scheme allows the network server to manage the data rate for each connected device, which allows the LoRaWAN connections to decide between payload and range dependent on local radio conditions.

The LoRaWAN network topology has three classes of endnode device. Class A are bidirectional devices, which have a scheduled uplink transmission window and two, short downlink receive windows. Class B devices have additional, scheduled downlink windows and Class C devices have open receive windows. Security for LoRaWAN includes the use of unique networks and device-encryption keys.

Name of Standard	LoRaWAN
Frequency band	433/868/780/915MHz ISM
Channel width	EU: 8x125kHz, US 64x125kHz/8x125kHz
	Modulation: Chirp Spread Spectrum
Range	2-5k (urban), 15k (rural)
Endnode transmit power	EU:<+14dBm
	US:<+27dBm
Packet size	Defined by User
Uplink data rate	EU: 300 bps to 50 kbps
	US:900-100kbps
Downlink data rate	EU: 300 bps to 50 kbps
	US:900-100kbps
Devices per access point	Uplink:>1M
	Downlink:<100k
Topology	Star on Star
Endnode roaming allowed	Yes

nWave

The nWave technology is an ultra narrow band (UNB) star topology radio technology and communications scheme. The nWave company offers radio modules, modems, and base stations for developers building their own private networks.

Name of Standard	nWave
Frequency band	Sub-GHz ISM
Channel width	Ultra narrow band
Range	10km (urban), 20-30km (rural)
Endnode transmit power	25-100 mW
Packet size	12 byte header, 2-20 byte payload
Uplink data rate	100 bps
Downlink data rate	--
Devices per access point	1M
Topology	Star
Endnode roaming allowed	Yes

Dash7

The Dash7 protocol defines communications between endnodes, sub-controllers, and gateways. In a typical network topology, endnodes communicate via sub-controller to gateways. This type of configuration requires that the endnodes periodically wake up and scan for any commands from the sub-controllers. However, if the endnode itself has a message to send it first must request the sub-controller to initiate communications with the endnode before it can send its own message. Subsequently, the sub-controller will then relay the messages to the gateway for transit to the Internet. Endnodes can communicate directly with other endnodes. The system communicates with the endnodes using asynchronous communications in a command/response format. The system can also use multi-cast queries to send commands to endnodes and have them respond only if they are a group member; for instance, if they have a temperature sensor or meet a specific operational condition such as endnodes that have a temperature sensor and have a reading greater than 10c and less than 13c.

Commands and responses are encrypted using AES-128 encryption but there is also a "stealth mode" whereby endnodes respond only to pre-authenticated devices.

Name of Standard	Dash7 Alliance Protocol 1.0
Frequency band	433, 868, 915MHz ISM/SRD
Channel width	25KHz or 200KHz
Range	0 – 5 km
Endnode transmit power	Depending on FCC/ETSI regulations
Packet size	256 bytes max/packet
Uplink data rate	9.6kb/s, 55.55 kbps, or 166.667 kb/s
Downlink data rate	9.6 kb/s, 55.55 kbps, or 166.667 kb/s
Devices per access point	N/A (connectionless communication)
Topology	node-to-node, star, tree
Endnode roaming allowed	Yes

Ingénue RPMA

The RPMA LP-WAN provides developers with transceiver modules that can connect to a network of access points, which form a global network that Ingénue and its partners are constructing. The transceivers in the endnodes connect to local access points that then forward messages from the endnodes to the user's IT system. However you are not required to use the Ingénue network, as it is also possible to purchase the access point devices and network appliances in order to build private networks.

The way the technology works is that the RPMA (random phase multiple access) transceivers and access points work together to control and optimize dependent of RF conditions the capacity, data rate, and range of their communications.

With regard to message privacy and security, Ingénue's RPMA includes two-way authentication, 256-bit encryption for confidentiality, and a 16-byte hash to protect the integrity of the messages.

Chapter 11 | IIoT WAN Technologies and Protocols

Name of Standard	Ingénue RPMA
Frequency band	2.4GHz ISM
Channel width	1MHz channels (40 channels available in 2.4GHz band)
Range	2000 miles
Endnode transmit power	20 dam maximum
Packet size	6 bytes to 10 Kbytes
Uplink data rate	AP aggregates to 624Kbps per sector (assumes 8-channel access point)
Downlink data rate	AP aggregates to 156Kbps per sector (assumes 8-channel access point)
Devices per access point	Up to 384,000 per sector
Topology	Typically star
Endnode roaming allowed	Yes

Low Power Wi-Fi

With Wi-Fi so ubiquitous in business and consumer IoT applications, the IEEE is working to expand the successful technology to low-power wide-area networking applications. The approach the IEEE is taking is to modify the PHY and MAC layers to enable low-power support for IoT applications. The standard is still under development, with initial release targeted for 2016.

Name of Standard	IEEE P802.11ah (Low-Power Wi-Fi)
Frequency band	License-exempt bands below 1GHz, excluding the TV whitespaces
Channel width	1/2/4/8/16MHz
Range	Up to 1Km (outdoor)
Endnode transmit power	Dependent on regional regulations (from 1mW to 1W)
Packet size	Up to 7,991 bytes (w/o aggregation), up to 65,535 Bytes (with aggregation)
Uplink data rate	150Kbps ~ 346.666Mbps
Downlink data rate	150Kbps ~ 346.666Mbps
Devices per access point	8191
Topology	Star, Tree
Endnode roaming allowed	Allowed by other IEEE 802.11 amendments (e.g., IEEE 802.11r)

LTE Category-M

The 3GPP is in the midst of defining a new release for LTE cellular technology that will define a Category-M device class targeting IoT applications. The standard is still being defined.

Name of Standard	LTE-Cat M*
Frequency band	Cellular
Channel width	1.4MHz
Range	2.5- 5km
Endnode transmit power	100 mW
Packet size	~100 - 1,000 bytes typical
Uplink data rate	~200kbps
Downlink data rate	~200kbps
Devices per access point	20k+
Topology	Star
Endnode roaming allowed	Yes

Weightless

Weightless is a collection of three LP-WAN standards under the control of the Weightless SIG (see http://www.weightless.org/). The original Weightless-W was planned to use the television whitespace frequency band for the wireless link. The proposed packet size and data rates are flexible, depending on user's need and link budget. However, Weightless-W is now on hold due to legal problems with the usage of the television whitespace frequency.

The Weightless-N standard is based on nWave's ultra narrow band LP-WAN technology and targets low-cost applications needing only unidirectional data transmission. An interesting feature of Weightless-N base stations is that they can operate in the same area as base stations from different service providers and still manage to interoperate. It achieves this because each base station queries a central database to determine which network an endnode is associated. Live deployments have begun in London and other European cities.

Chapter 11 | IIoT WAN Technologies and Protocols

The Weightless-P standard is under development and scheduled for release in late 2015 with hardware available in early 2016. The Weightless-P link is based on networking technology originally developed by M2 Communication and will provide fully acknowledged bidirectional communications. Weightless P shares its MAC layer with Weightless-W, and will support fast network acquisition with hand-over of roaming endnode devices across base stations.

Name of Standard	Weightless		
	-W	-N	-P
Frequency band	TV whitespace (400-800MHz)	Sub-GHZ ISM	Sub-GHZ ISM
Channel width	5MHz	Ultra narrow band (200Hz)	12.5kHz
Range	5km (urban)	3km (urban)	2km (urban)
Endnode transmit power	17 dBm	17 dBm	17 dBm
Packet size	10 byte min	Up to 20 bytes	10 byte min
Uplink data rate	1 kbps to 10 Mbps	100bps	200 bps to 100 kbps
Downlink data rate	Same	No downlink	Same
Devices per access point	Unlimited	Unlimited	Unlimited
Topology	Star	Star	Star
Endnode roaming allowed	Yes	Yes	Yes

Millimeter Radio

Recently there has been a lot of interest in the unlicensed 60Ghz wave band in the millimeter range. One company, Starry, was heavily involved in Weightless-W TV whitespace band and has launched a broadband Internet service in urban areas on this 60GHz frequency. Starry believes that it can utilize this underused high-frequency band to deliver high-speed broadband Internet in urban areas. There are pros and cons with all radio transmissions, but one fact is immutable—you can have high bandwidth directly related to high frequency and you can have range, for example the distance you can reliably transmit, but you cannot have both. At 60GHz there is undoubtedly the potential to send high throughput but it will be at sort distances—less than 1km—and it will need to be clear line of sight. The problem with 60GHz millimeter radio waves is they cannot penetrate an interior wall, let alone a building that blocks the line of sight.

Facebook is also pursuing this technology. They are trying to patent a mesh technology that they can deploy in remote areas to bring broadband Internet to rural villages in India as part of their Ineternet.org project. Facebook like Starry is very confident in the technology working in both rural and urban environments. However it should be noted that Facebook intends using a mesh network whereas Starry intends to use a point-to-point topology.

The problem is that there is another immutable law of radio—the higher the frequency, the more susceptible it is to rain fade and atmospheric conditions. Therefore, even to survive light rain requires high power and that alone rules it out for most if not all IoT use-cases.

References

http://www.ibm.com/internet-of-things/

http://electronicdesign.com/iot/understanding-protocols-behind-internet-things

CHAPTER 12

Securing the Industrial Internet

Security is one of the biggest inhibitors to adoption of the Industrial Internet, the deep-seated fear of opening up industrial processes to potential disruption or the loss of critical business secrets to the Internet strikes deep. Traditionally, industrial networks have managed to remain immune to most of the scourges of the Internet such as viruses, worms, Trojans and DDos attacks, simply because their architecture and protocols are so different from IT enterprises and consumer computer devices.

Rarely do industrial systems run on Windows or Linux, instead the vast majority run on small proprietary operating systems connected over non-IP protocols and serial bus topologies. Additionally, many of these networks have air-gaps between the support departments, such as finance, sales, customer support, and IT's IP networks, which provide for a degree of isolation. Even if there were a direct network connection to the Internet, a simple VPN pipe would be all that would be required to interconnect remote facilities for M2M inter-communication any other non-M2M traffic would travel across the IT Internet gateway. It is because of these characteristics of industrial networks that they have remained somewhat unsusceptible to many of the security issues related to the Internet.

However, it appears that common belief may actually be mistaken, as some security researchers claim that industrial exploits have been deliberately kept secret. Indeed, ICS-CERT, an industry watchdog, claims that of the 245 suspected exploits they investigated in 2014, many of them could not be determined due to lack of information. Additionally, in a Black Hat convention in 2015, a researcher from Hamburg University of Technology, Marina Krytofil, claimed that the hacking of industrial facilities for extortion was one of the biggest untold security stories. Furthermore, she claimed that large-scale hacking for extortion had been prevalent in the industry since 2006, with hackers targeting large-scale utility companies, but nearly ten years on there was little information on how they were doing it.

Marina Krytofil's research is interesting as it highlights the motives behind the hacking attacks as being persistent financial gain via extortion rather than physical damage and hence the companies being unwilling to report the exploits. This is interesting as it indicates that the hackers have a persistent attack vector that can remain undetected over long periods. Furthermore, it would also suggest that the hackers have detailed industrial system control and process expertise in order to take over a system and then to manipulate a process to achieve a desired outcome—to demonstrate control, but not damage a process—while at the same time covering their tracks to avoid detection.

Companies are reluctant to announce these security breaches due to fear of loss of reputation and damage to their brand. However, things may be changing as of 2016, because security researchers are seeing increasing numbers of security exploits aimed at industrial targets. The latest of which brought about a power outage across the Ukraine. This was the first coordinated attack believed to have been directly responsible for a power outage. However, the Ukraine power utility company exploit was by no means the first of its kind; it is just the first to have sabotaged production, other more notable attacks, such as the Dragonfly exploits on U.S. And European power companies had industrial espionage as the key motive.

Dragonfly, a group believed to be made up of Russian hackers, has been operating since at least 2011. Dragonfly first came to prominence and notoriety when it began targeting defense and aviation companies in the United States and Canada. However, a shift in strategy in 2013 saw the group suddenly transfer its focus into targeting U.S. And European energy firms.

Dragonfly's mode of operation is now well documented and understood by security researchers as they gain entry through these methods. Initially they try the method of least resistance and go for a campaign of spear phishing e-mails delivering malware. Dragonfly also uses sophisticated watering hole attacks that redirect visitors to energy industry-related web sites hosting an exploit kit, which will infect the victim's browser using an embedded JavaScript in the host web server's web page. An alternative attack vector is through infecting legitimate software made available for download from three different respectable ICS (industrial control systems) equipment manufacturers.

Presently, Dragonfly's main motive seems to be cyber-espionage. However, the sheer amount of intelligence that Dragonfly has reaped over the years from European and U.S. energy companies means it could be in a position to do a lot of damage. Despite this, though, there is no evidence of any malicious intention such as sabotage so it isn't clear what Dragonfly's commercial or ideological motivation is, if it is not state-sponsored.

Not all attacks are malicious or motivated by financial or personal gain; some are just done for the sake of it. For example, the Godzilla Attack is a perfect example of a potential industrial Internet-type security exploit on a smart city traffic system where the attacker gained access to the traffic lights and digital overhead messaging system.

In May 2014, the overhead traffic signs on San Francisco's Van Ness Avenue were photographed by a passerby showing the flashing traffic warning "Godzilla Attack! Turn Back".

This humorous attack highlights just one of many traffic light hacks that have come about, but it could have serious implications for traffic management systems, in the context of smart cities. If traffic lights are vulnerable and an attacker can easily exploit them, it is only a matter of time before a malicious attacker can take advantage of the vulnerability to bring about massive traffic congestion or create potentially dangerous traffic flows that could lead to injury or even death.

It is not just smart cities, and industrial companies and facilities that are appetizing targets. Sometimes the products are so intelligent that they themselves become the object of research and potential exploitation, as both Chrysler and Boeing discovered with the now-famous Jeep and aircraft hacks. Both of these celebrated attacks came about on systems thought to be unsusceptible to cyber-attacks. The manufacturer's belief of invulnerability came about predominantly based on the technical assurance that both the jeep and the Boeing aircraft had isolated control systems. Both use the vehicle standard CAN (control area network) bus for interconnecting in-car and in-flight systems and modules, that are not connected to any IP networks. Now that might well have been true previously, but with the introduction of IP networks for infotainment systems that utilized Wi-Fi access, this was no longer the case. By gaining access to the Wi-Fi information systems, researchers in the Jeep hack were able to discover a device that connected to both the CAN and Infotainment IP network—the V850 controller.

The V850 controller provided only read-only access to the CAN network, hence the manufacturer's confidence in the isolation of the CAN bus. However, being a computer capable device, the researchers were able to change the way it worked through programming and they reconfigured the device using a firmware upgrade to provide read/write access to the CAN bus. The result was spectacular in that they could now control every aspect of the Jeep, as the CAN bus interconnects every component of the vehicle from the engine to the transmission and all the actuators and sensors, controlling everything from the wing mirrors to the drive chain.

With control over the CAN bus the researchers were able to control the steering wheel, engine, transmission, braking system, and even things like the windscreen wipers, air conditioner, door locks, and so on. Moreover, they were able to control all these components remotely, over the Sprint cellular network.

In the Boeing example, researchers also claimed to have gained control of the CAN bus and be able to manipulate the critical in-flight systems such as the engines, wing flaps, tail rudder, and navigation systems. Boeing strenuously denied this, fobbing of the reports as being hacks based on a flight simulator and not a real aircraft, claiming again that there was isolation between the IP infotainment Wi-Fi systems, and the critical in-flight CAN bus network. However, it is notable that in 2013 and 2014 Boeing did request permission to modify the 747—the model researchers claimed to have successfully hacked—stating that the modification was required to separate the IFE (in-flight entertainment) from critical system in the aircraft.

> "The architecture and network configuration may allow the exploitation of network security vulnerabilities resulting in intentional or unintentional destruction, disruption, degradation, or exploitation of data, systems, and networks critical to the safety and maintenance of the airplane."
>
> —FAA (https://www.regulations.gov/document?D=FAA-2014-0301-0001)

Researchers in the Boeing project were not malicious and had brought the potential vulnerabilities to the manufacturer's attentions; however not everyone is so well intentioned, and as a confirmed attack on a German steel mill highlighted, the intention of the attacker can simply be to cause physical damage.

In 2014, a concerted cyber-attack was directed at a German steel mill via its commercial IT network. Once access had been obtained to the IT network, gained by sophisticated phishing and social engineering scams, the attackers strived to gain access to the ICS (industrial control systems) network. Their strategy was to use the IT network as a jump pad so the attackers found an undisclosed method to gain access and control of the ICS network. The damage they caused resulted in the altering of a furnace's internal characteristics causing it to stop responding to control systems orders to shut down. This resulted in extensive damage, as well as reputational, productivity, and revenue loss.

Security in Manufacturing

Security is a major issue with Industrial Internet applications as IOT (industrial operations technology) environments are not like IT networks; there are rarely firewalls and intrusion detection appliances because there was simply little requirement for them on the IOT's flat networks as they were isolated LANs with no Internet connection. Of course with the introduction of IIoT, that splendid isolation no longer exists and securing systems and applications is now a major concern.

One of the issues is that IOT networks differ greatly from traditional IT, for example, they traditionally used non-IP protocols for M2M and machine to PLC (programmable logic controller) communications. In addition, IOT networks were often built on field bus type topologies, or ring topologies using protocols such as Modbus, which daisy chained machines and devises using point-to-point serial RS232/422 or RS485 cabling. Protocols such as Modbus RTU were very efficient as they worked in a master/slave mode of operation and transmitted asynchronous unscheduled messages at 9.6Kbps or at 115Kbps. Modbus could also broadcast to up to 247 slaves simultaneously in a bus topology. This is of course archaic in an IT context as their networks are based on 100Gbps Ethernet core switches and 10Gbps servers, with 1Gpbs at the user's desk. Even in wireless deployments, IT delivers 10's of megabytes of bandwidth over the air. However, this is one of the main issues, in that IOT systems do not require high throughput. They communicated through M2M or PLCs very efficiently using small messages over a very locally connected network. Performance was not the requirement, or not the overriding one it is in IT scenarios. In operation technology, which differs from it (information technology) in so much as OT is related to industrial networks where availability is king.

System availability is hugely important in industrial OT environments, and it is viewed in a way that OT technicians and engineers feel IT does not really understand. In industrial OT, there cannot be any downtime to reboot servers, which is why Windows Server was rarely deployed in industrial, as it did not satisfy OTs demand for availability. Rebooting a server perhaps every couple of months is not something OT would consider acceptable. Similarly, Ethernet was deemed to lightweight and fragile—commercial versions with plastic RJ45 connectors were susceptible to heavy vibration and excessive heat. In addition, the early Ethernet protocol was non-deterministic in its half-duplex basic mode.

The rejection of Ethernet and Windows in industrial OT networks highlights another major difference between OT and IT systems. Reading the last paragraph you would be excused for thinking OT was living in some timewarp, after all, the days of Ethernet being half-duplex over wire networks are long gone. However, what you have to understand is OT systems are hugely expensive and are purchased on a 20-year lifecycle. So when you survey all the machinery in a modern factory you are probably getting insight into the technology that was available and the technical decisions that were made perhaps as long as 10 to 15 years ago. In relation to their original design and purpose, they work perfectly well. Despite this OT systems were not standing still; they were adapting and gradually new field bus technologies started to replace older protocols or they were, as in the case of Modbus, converted to handle IP traffic. The trend to use IP and Ethernet, albeit in an industrial version Ethernet/IP (the IP stands for Industrial Protocol), is accelerating through IP communication protocols such as Profinet and Ethernet/IP.

The problem is that technology is changing rapidly and some old and new technologies have advanced in recent years beyond all predictions back in the early 2000s. These advancements, such as in sensor technologies, wireless radio technology, and Big Data analytics, among many others have enabled the concept of the Industrial Internet and all that it promises and threatens.

An important point is the comparison between IT networks back in the early 1990s, before the Internet and when networking was at a rudimentary level. For example, when PCs were connected to Windows NT server running over Ethernet and NetBIOS. TCP/IP was available, it had been around for years but was unnecessary as the Internet was still in its infancy, in comparison to what it was to become. Suddenly, the Internet stopped being a vague hyped-up concept promising the world, and became a reality. By 1995 web browsers such as Netscape made it accessible to the public.

The point is that prior to that point, there had been computer viruses, worms, and the like, but nothing like the scale of what was to come with the advent of the Internet. Once the Internet became ubiquitous in the business and the home and e-mail became the de-facto form of communication, virus manifestations rocketed. IT was ill prepared for such attacks.

As a result of the proliferation of Internet connected networks, computers started to be hacked remotely, and IT appeared defenseless against these violations as the concept of an IT security department was still years away. However, these threats escalated over the mid-to-late 1990s and reach their zenith with the mass-hysteria conjured up over the dreaded Millennium Bug (Y2K). This specific threat was not a virus, a worm, or even a deliberate hack. It was simply the recognition of the fact that programmers in the 70s and 80s used two digits to represent the year in their code. For example, the 19 in 1976 programmers took for granted, so they represented 1976 simply as 76 and they never thought twice about the date field format changing in the year 2000.

The Y2K bug, as it became popularly known, caused near panic within businesses and governments, with mass scale Y2K initiative set up an all software and applications had to be checked and verified as being Y2K compliant; this was a truly massive task. The fear within industry is connecting systems and networks that have never previously been connected to the Internet or even the outside world will bring down on industry similar woes as befell IT back in 2000.

Fortunately for industry, IT has learned tremendously over the last two decades how to protect code, applications, servers, and networks from the threats on the Internet. However, OT networks are not IT networks and as we have seen, they have different topologies, components, protocols and purposes. On the other hand, IT does have in many cases limited knowledge of the requirements of OT systems, including how they differ and the possible consequences of ill-advised action.

Consequently, we need to look at a typical OT network in manufacturing to see what exactly is different and how IT security techniques and knowledge can be applied to protect these networks from the threats that we know exist today.

From a security prospective, it might be tempting to think we can just transfer our knowledge from IT to OT, as OT is often considered simply a specialization or a subset to IT. However, that approach is fundamentally flawed as the OT or ICS (industrial control system) environment is very different from the IT enterprise. IT is deemed to be all the systems, applications, people, and networks required to manage information technology. ICS, on the other hand, is all the control systems, PLCs, DMI, people, and networks responsible for operational and production tasks.

One of the components that defines these architectures, whether it is IT or OT, is people and it is important to remember that OT and IT staff have very different perceptions of what is fundamentally important to the successful operations of their respective domains, including security.

As an example IT security staff would consider the fundamental concepts of IT security, CIA, confidentiality, integrity and availability to be the foundations of enterprise security. The OT support staff, however, although agreeing in principle, would stress that availability should be given the greatest weighting when evaluating importance and, therefore, the greater focus. Now to the actual depth of the divide between the parties, consider some common IT network tools for troubleshooting an administering a LAN. IT support staff commonly ping devices on the network, often using a continuous ping using the -t switch. Similarly, when mapping or for network discovery, they would think nothing of running a scan on the local network, indeed if not falling under a best practice it is certainly a common practice.

However, both these techniques would be considered abhorrent in OT or ICS environments, because the tools used, PING and ICMP for discovery scans, interrupts the end-node (computer or IP enabled device) from doing is current task, by effectively pulling rank, and telling it, stop what you are doing and service my request. For IT that is not a problem as it would have minimal effect on IT applications where timing isn't critical, as they deal with system to human communications, where a few milliseconds delay is undetectable. However, in the time-critical OT and ICS domains, these tools could potentially be detrimental to performance. Consequently, the analysis and troubleshooting tools and their method of operation must be carefully analyzed before they are used in OT and ICS environments.

However, there is a far more important issue that needs to be resolved when attempting to merge OT with IT and this is something that IT rarely acknowledges, as it is alien to their normal working environment. That missing factor is safety, which does not feature in the three tenets of IT security—confidentiality, integrity, and availability—but is hugely important in OT environments, as it can prove catastrophic to employee safety. Health and safety in industrial environments is paramount as they tend to be very dangerous environments, and bringing IT technicians into an OT environment without comprehensive training is a recipe for disaster.

It is consequently vital that a security person realizes that in hazardous environments, safety and availability trumps the fundamental tenets of IT security, confidentiality, integrity, and availability. In the perspective of OT, they would define the fundamental security tenets to be safety, availability, confidentiality, and integrity.

With this understanding in mind, it is crucial when considering security in an OT environment, that the IT security person first know the hazards and nature of the OT environment in which they will be working. Safety training should be addressed through mandatory training before anyone is even admitted to the site. In practice, this is not always possible, after all, you cannot always mitigate human thoughtlessness. Despite that, OT security must have a far greater emphasis on physical and behavioral security rather than on the bits and bytes crossing the wires.

Another significant problem when trying to have IT secure the OT domain is lack of basic understanding of the technologies and protocols. For example, what are PLCs, DCSs, and HMIs? In addition, what are those strange non-IT protocols in use in M2M communications?

To illustrate the issue, a typical OT network would look something like Figure 12-1.

Industry 4.0

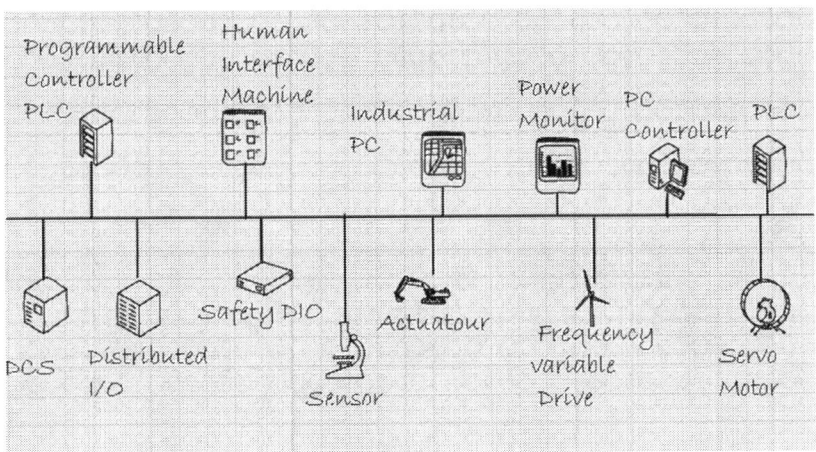

Figure 12-1. Typical OT manufacturing network

Figure 12-1 shows a typical OT and ICS network topology. There are references to non IT systems such as PCL, DCS, HMI, field bus, DCP3, Modbus, Ethernet/IP and Profinet, which unless the IT security person has a background in industrial control systems will leave them baffled. So what are these entities?

OT and ICS networks consist of PCL and DCS, a PLC (program control logic) is used to automate processes by running a program, which has been created in software to replicate the actions required, in a step basis, to complete a task. PLCs are digital and they provide the necessary digital outputs required as input by microprocessors to perform a function, such as to turn on or off a motor. PCLs also take digital input, and this requires that sensor data, which is often analogue in nature. This must be converted by an analogue-to-digital convertor (ADC) before being presented to the software running on the PLC. A PLC runs specific control logic software for a machine and process so there is often a tight correlation between a PLC and a machine and the process it runs. Consequently, there are many PLCs in a traditional OT and ICS environment, such as a manufacturing factory floor, and this can cause a huge proliferation of wiring and the consequential costs in provisioning and maintaining the wiring burden.

A DCS on the other hand is also used to automate systems. However, a DCS is a distributed control system that is often deployed in the automation of continuous and large-scale processes. Furthermore, an HMI (human machine interface) is a control system that allows a human operator to interact in real time with the machine and control the process running on the machine. HMI systems are typically in the control room and operated by machine and process operators. However, very few IT security personnel would be familiar with these terms, let alone know how to secure them.

Chapter 12 | Securing the Industrial Internet

HMI, PLC, and DCS make up the systems in an industrial network, and there are only a handful of automation system vendors, such as Siemens, Rockwell, General Electric, and ABB, among others. This is a reassuring point as these automation vendors, though not just the ones listed make the vast amount of systems, sensors, and controllers deployed in production OT networks. This is important as we can more readily research and request technical specifications and security documentations and best practices from the relevant vendor. The reason this is so important in OT networks is that although some software is supplied for Windows and Linux deployments, it is actually quite rare. The vast majority of devices and sensors run on proprietary lightweight but real-time operating systems, which can deliver the performance and deterministic responses required in time-critical process control.

PLCs and DCS

If we compare the OT network topology diagram with an IT security topology, shown in Figure 12-2, we can see the major functional differences. Therefore, we need to ask ourselves, what are all these areas?

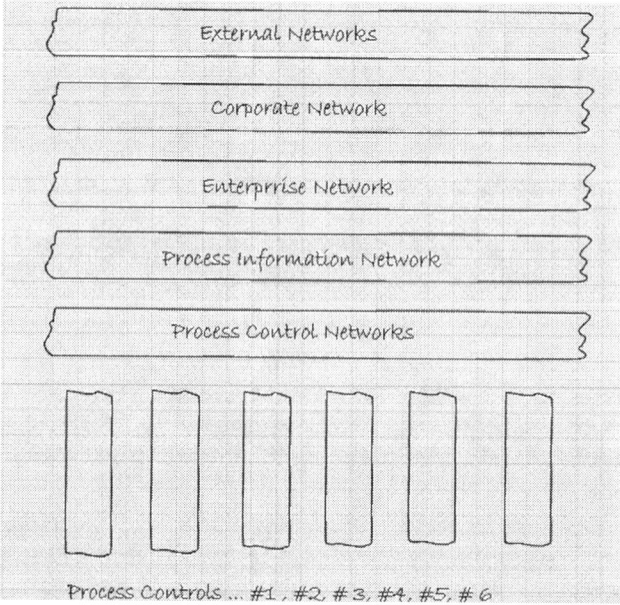

Figure 12-2. OT versus IT security domains

Securing the OT

The first thing we need to consider is that like early IT enterprise networks back in the early 1990s, OT networks may support several different protocols and standards, unlike modern IP there is no standardization on Ethernet and IP. Therefore, different parts of the network could feasibly support different network types, such as Modbus, Ethernet/IP, Profibus, and Profinet, as well as a mixture of wiring media and standards, such as fiber optics, copper cables, or wireless for Ethernet 10/100 Mbps or Gigabit Ethernet. OT networks evolve along these lines, as there was no industry standard and with the costs and lifespan of equipment manufacturers tended to supply and install equipment that was favored at the time. Lately, these has been an industry-wide move to IP and Ethernet; however, even then there has not been a standard achieved. The United States has a preference for Ethernet/IP, Europe favors Profinet, and Asia use a mixture of the two.

In Figure 12-2, you can see the bus structure of the network and each area of production or process is defined but without any of the data segregation, routers, or firewalls common in an IT network topology. The common IT practice is to segregate each area preventing unauthorized or irrelevant traffic from entering a specific area or zone, but consider the latency, jitter, or packet loss this might introduce.

Network Level: Potential Security Issues

Considering the network topology is a good place to start as the potential to contain and isolate a security threat is enhanced. The bus structure traditionally favored in industry, due to its flexibility—new systems can be simply added without network redesign or disruption—is also prone to security vulnerabilities such as worm infestations. The problem is that a flat network, which is what a bus is, can be, with limited hosts, a highly efficient communications network, which makes it so popular. Ethernet, is a prime example of fast switching at the data packet level, it is fast and simple to install and administer. However, Ethernet-switched networks are great for small LANs but they do not scale well, due primarily to its broadcast nature. Isolation of network zones can be achieved by using VLANs, which virtualizes the underlying network infrastructure by creating virtual LANs through tagging packets from a predetermined source travelling to a destination with an individual VLAN identifier.

VLANs reduce the size of broadcast domains, thus improving efficiency and allowing flat switched networks to grow. However, flat switched networks have a number of issues. The most important being is that the topology is very important. In an industrial field bus topology, such as Profinet, this is not typically an issue, indeed working at the MAC address level provides for increased and deterministic behavior, so it is actually a good thing. However, as IP is introduced into the mix, it may become a serious issue as loop detection protocols are then required. This is because a loop in the network topology can have serious consequences for switches knowing what port to send out packets.

The most common of the loop detection protocols used in IT environments was spanning tree. However, spanning tree would be unacceptable in an industrial production environment because of the time it takes to detect loops and run the SP algorithm before restoring network connectivity, which can be even in fast spanning tree counted in seconds of traffic-flow disruption.

In IT enterprises they tend to default to hierarchal network designs such as those favored in IT traditional data center networks—prior to SD—and this topology relies on IP subnets to provide segregation and the mechanisms for border access control at routing devices between segments or security zones. Access control through subnet-centric access lists in IP works well, especially in IT enterprise networks, even though it does add some latency (though that latency is undetectable in IT applications). The problem lies with configuring the routing devices; in small environments they could be statically configured with hard-coded routing tables that explicitly state where and through which of the routers interfaces the destination can be reached. This is efficient in operation but prone to errors during configuration, it also is not scalable; as the network grows, it becomes increasingly difficult to configure and maintain.

The alternative to static routes is routing protocols, such as RIPv2, EIGRP, ISIS, and OSPF. However these are at default settings also unsuitable for industrial OT environments. The reason for this is that these routing protocols were designed a long time ago where 20 to 30 second convergence times weren't an issue. Even today with faster configuration options for these routing protocols, network convergence still fail to deliver the micro-second requirements for sensors in the OT and ICS world.

The question, then is, what flavor of Ethernet do OT environments in time-critical M2M environments use? The simple answer is industrial Ethernet, such as Ethernet/IP, which can be Devicenet over Ethernet, or Profinet, which comes from the same family as the established field bus protocol, Profibus, though they are radically different so should not be confused.

To illustrate how industrial Ethernet differs from the IT and data communications standard of Ethernet, we need to investigate how Profinet works and how it mitigates the failings of standard Ethernet, as well as why it is so advantageous to OT.

System Level: Potential Security Issues

At the system level, PLC and DCS are the vulnerable targets and potentially the target for malicious attacks as both run the software that controls the underlying logic that runs in their respective automation processes. A malicious attacker may well wish to gain control of the software and shut down systems or reap other forms of havoc. Although industrial systems have been relatively unaffected by outside attacks compare to IT systems, it is a worrying upward trend. Indeed, the number of reported attacks on industrial targets has risen considerably in the last few years. ISC-CERT has been monitoring the attack landscape since 2010 and in five years has reported around 800 advisories related to exploits, vulnerabilities, and other security concerns. This of course is miniscule concerned to the level of exploits targeting Android mobile devices for example, or the Windows OS on PCs and servers. However, it is certainly not something that can be ignored, as the potential consequences of a malicious attack on an industrial complex, such as a nuclear power station, oil rig, or a petrochemical refinery could be devastating with catastrophic loss of life.

ICS-CERT dealt with 245-suspected security exploits during 2014 alone; however these are only the reported issues, and they found disconcertingly, that the root cause of the exploit could not be determined, hence suspected security exploits. The reason for ICS-CERT being unable to determine the attack-vector or the source of the attack is that the OT systems suspected of being compromised generally did not have detection and logging capabilities. Not all OT and ICS systems on an OT network have inherent to their design the capabilities to identify, authenticate, authorize, and apply access rules. Vendors are addressing these issues in their latest products, but remember the long lifecycle of products already deployed.

However, it is critical that to secure any network you know what you are protecting. For example, in a manufacturing plant you must be able to identify and manage every sensor or node in the network. Similarly, how can you ensure that every node or sensor in your smart city network is one of yours and not some rouge device planted by an intruder? In order to protect the network, we need to be able to identify, authenticate, and manage every node on the network, and for that we need IAM (identification and access management).

Identity Access Management

In enterprise IT, the requirement for IAM (identity Access Management) became a priority due to the proliferation of consumer devices into the workplace encouraged by initiatives such as BYOD (bring your own device). Suddenly, with the huge response and uptake of BYOD, IT had a real problem on their hands, as they could not identify all these employee devices, smartphones, tablets, laptops, and even wearable IoT devices that accessed their networks. However, IT used mobile phone technology to identify and enforce policy and security templates onto massive online social media systems such as Google search, maps, and even web-scale applications and platforms such as Facebook.

Chapter 12 | Securing the Industrial Internet

In the previous example, we considered manufacturing from the myriad of industrial examples where security affects other streams of industry such as aviation, nuclear power, electricity grids, health care, and vehicle automation.

Consider the development of the massive airliner, the Airbus A380, as an illustration of how even the most financially endowed manufacturing project could be forced to mix and match protocols and standards. The Airbus A380 example is important as the project budget was billions of U.S. dollars and the price of a single aircraft was $428 million. Yet, here is the strange thing—the designers, presumably with years to design and plan ahead, and with sufficient budget to issue contracts to perhaps redesign or reimagine required components, still ended up launching an airliner with multiple networks.

The A380 launched with so-called state of the art infotainment systems, in so much as it supported IP and Ethernet for video, music and in-flight Wi-Fi. However, for the flight control systems it stayed with the traditional CAN bus as the physical topology as it was industry proven, a common interface, and even though its communication throughput was limited to 125-500Kbps that was sufficient, as performance was deterministic. However, the A380 also had a legacy bus network installed in the cockpit to support the VHF radio equipment. The reason for this was simply all VHF radio equipment comes with standard VHF interfaces—much like IT equipment comes with common Ethernet connections. The result was that Airbus designers made the pragmatic decision to use standard equipment and interfaces even though it might add maintenance complexity, it would certainly cost less and would accelerate development. Another interesting design element in the A380 and in high-performance cars and aircraft is that the human machine interface is almost always analogue, with dashboards displaying information using vague readings displayed as dials and needles rather than accurate digital numeric displays. The reason for this is humans understand the analogue world quicker than they do the digital world. A driver or pilot can glance at an instrument panel, with perhaps dozens of dials, and immediately notice that one dial is showing in the red, the extreme of its limit. With digital displays, the driver or pilot would have to read and then compute the digital information, for example oil pressure, and then mentally input this data and compare it against his understanding of the safe limits, which takes time and knowledge. Consequently, the visualization of data in human machine interfaces tends still to be displayed graphically mimicking analogue displays, in less accurate but quicker-to-assimilate analogue format.

Aircrafts and high-value cars that have adopted IP and Ethernet in place of the tried and tested CAN bus that served both so well for decades, also have the problem of being vulnerable to advanced malware and exploits. As an example, Modbus, Devicenet, Profibus, CAN bus, and other industrial network topologies are relatively immune to most modern malware, due to their inaccessibility—no remote Internet connection and unique protocols—so cyber-attacks on these serial bus technologies are rare. However, that is not the case with Ethernet and IP. Although some will argue that Ethernet and IP-enabled control systems have the advantage of a vast body of knowledge has been built up over two decades of computer attacks, with databases full of recorded vulnerabilities, known exploits, flaws and mitigation techniques, they are considered battle hardened and secure. However, that works both ways—Ethernet and IP systems that are not state of the art, or protected by diligent security measures, are open to a vast battalion of exploits and attacks, developed and tested over many years.

It is important to remember that OT environments will contain protocols and technologies that are simply unknown or considered archaic in the enterprise. This is due in some part to the long lifespan of industrial assets, wiring asset cost reduction, efficiency, and simplicity and in no small part to the relative immunity from common Internet threats such as malware. Consequently, although Ethernet and IP may be making inroads into OT environments, field bus technologies still dominate the industrial landscape when it comes to communication protocols—and will do so for many years.

References

https://www.blackhat.com/us-15/speakers/Marina-Krotofil.html

https://blog.kaspersky.com/blackhat-jeep-cherokee-hack-explained/9493/

http://www.wired.com/2015/05/possible-passengers-hack-commercial-aircraft/

CHAPTER 13

Introducing Industry 4.0

Industry 4.0's provenance lies in the powerhouse of German manufacturing. However the conceptual idea has since been widely adopted by other industrial nations within the European Union, and further afield in China, India, and other Asian countries. The name Industry 4.0 refers to the forth industrial revolution, with the first three coming about through mechanization, electricity, and IT.

The forth industrial revolution, and hence the 4.0, will come about via the Internet of Things and the Internet of services becoming integrated with the manufacturing environment. However, all the benefits of previous revolutions in industry came about after the fact, whereas with the forth revolution we have a chance to proactively guide the way it transforms our world.

The vision of Industry 4.0 is that in the future, industrial businesses will build global networks to connect their machinery, factories, and warehousing facilities as cyber-physical systems, which will connect and control each other intelligently by sharing information that triggers actions. These cyber-physical systems will take the shape of smart factories, smart machines, smart storage facilities, and smart supply chains. This will bring about improvements in the industrial processes within manufacturing as a whole, through engineering, material usage, supply chains, and product lifecycle management. These is what we call the horizontal value chain, and the vision is that Industry 4.0 will deeply integrate with each stage in the horizontal value chain to provide tremendous improvements in the industrial process.

© Alasdair Gilchrist 2016
A. Gilchrist, *Industry 4.0*, DOI 10.1007/978-1-4842-2047-4_13

At the center of this vision will be the smart factory, which will alter the way production is performed, based on smart machines but also on smart products. It will not be just cyber-physical systems such as smart machinery that will be intelligent; the products being assembled will also have embedded intelligence so that they can be identified and located at all times throughout the manufacturing process. The miniaturization of RFID tags enables products to be intelligent and to know what they are, when they were manufactured, and crucially, what their current state is and the steps required to reach their desired state.

This requires that smart products know their own history and the future processes required to transform them into the complete product. This knowledge of the industrial manufacturing process is embedded within products and this will allow them to provide alternative routing in the production process. For example, the smart product will be capable of instructing the conveyor belt, which production line it should follow as it is aware of it current state, and the next production process it requires to step through to completion. Later, we will look at how that works in practice.

For now, though, we need to look at another key element in the Industry 4.0 vision, and that is the integration of the vertical manufacturing processes in the value chain. The vision held is that the embedded horizontal systems are integrated with the vertical business processes, (sales, logistics, and finance, among others) and associated IT systems. They will enable smart factories to control the end-to-end management of the entire manufacturing process from supply chain through to services and lifecycle management. This merging of the Operational Technology (OT) with Information Technology (IT) is not without its problems, as we have seen earlier when discussing the Industrial Internet. However, in the Industry 4.0 system, these entities will act as one.

Smart factories do not relate just to huge companies, indeed they are ideal for small- and medium-sized enterprises because of the flexibility that they provide. For example, control over the horizontal manufacturing process and smart products enables better decision-making and dynamic process control, as in the capability and flexibility to cater to last-minute design changes or to alter production to address a customer's preference in the products design. Furthermore, this dynamic process control enables small lot sizes, which are still profitable and accommodate individual custom orders. These dynamic business and engineering processes enable new ways of creating value and innovative business models.

In summary, Industry 4.0 will require the integration of CPS in manufacturing and logistics while introducing the Internet of Things and services in the manufacturing process. This will bring new ways to create value, business models, and downstream services for SME (small medium enterprises).

Defining Industry 4.0

If we look for a clear definition of Industry 4.0, it can prove to be quite elusive. As an example, here are three definitions:

- "The term Industry 4.0 stands for the fourth industrial revolution. Best understood as a new level of organization and control over the entire value chain of the lifecycle of products, it is geared towards increasingly individualized customer requirements. This cycle begins at the product idea, covers the order placement and extends through to development and manufacturing, all the way to the product delivery for the end customer, and concludes with recycling, encompassing all resultant services. The basis for the fourth industrial revolution is the availability of all relevant information in real time by connecting all instances involved in the value chain. The ability to derive the optimal value-added low at any time from the data is also vital. The connection of people, things and systems creates dynamic, self-organizing, real-time optimized value added connections within and across companies. These can be optimized according to different criteria such as costs, availability and consumption of resources." (PWC.DE)

- "A framework for Industry 4.0 depends on 1) the digitization and integration of the horizontal and vertical value-chains. 2) The digitization of products and services and 3) the introduction of innovated business models."

- "Industry 4.0 is a collective term for technologies and concepts of value-chain organizations. Within the modular structured smart factories of Industry 4.0, CPS monitor physical process, create a virtual copy of the physical world and makes decentralized decisions. Over the IoT, CPS communicates and cooperates with each other and humans in real time. Via, the IoS, both internal and cross-organizational services are offered and utilized by participants of the value chain."

If these definitions from well-reputed sources do not clarify what defines the Industry 4.0, perhaps we can look to practical examples.

The Industry 4.0 definition can be somewhat confusing; some will claim it to be "making the manufacturing industry fully computerized". Whereas, other may say it is a way to "make industrial production virtualized". However, the consensus appears to be "that it integrates horizontal and vertical channels". Either way, it is a huge incentive for businesses and manufacturers to keep up with the rapid pace of changes driven by the evolution of many enabling technologies.

Industry 4.0, like so many new technologies in the 21st Century, is not a new concept; it is more a rebirth of an older concept that is utilizing newly developed technology. To be precise, Industry 4.0 is essentially a revised approach to manufacturing that makes use of the latest technological inventions and innovations, particularly in merging operational and information and communication technology.

Industry 4.0 deploys the tools provided by the advancements in operational, communication, and information technology to increase the levels of automation and digitization of production, in manufacturing and industrial processes. The goal is to manage the entire value chain process, by improving efficiencies in the production process and coming up with products and services that are of superior quality. This vision follows the maxim of higher quality, not at the expense of lower price. This philosophy has produced the smart factory of the future, where efficiencies and costs improved and profits increased. This factory of the future is already here—as we will see later—to be one that operates with quiet efficiency, where all processes, driven by CPS and humans alike, are unlike any traditional factories, as they perform in almost sterile environments, cleanly, safely, reliably, and efficiently.

Why Industry 4.0 and Why Now?

The rise of the machine, heralded back in the 70s and 80s as the future of manufacturing and the solution to erratic humans on the production line, caused great worry. The concern was that machines—robots—would run our production, and, as it was initially successful in heavy industry, it resulted in the advent of automation in industry. Automation is rooted in the 80s, where the desire for efficiency in manufacturing resulted in the loss of many low-wage manual workers jobs, which was deemed the end for humans on the production line. It did not work out that way, although many workers did lose their jobs and livelihoods.

The rise of the machine and the robot came about through computers, IT, and semi-intelligent robots replacing many workers. However, this fourth industrial revolution is a transition to the digital transformation of the manufacturing industry—a merging of the physical and digital worlds—which holds other possibilities and does not necessarily mean downsizing.

Industry 4.0 has come about through several technological advances:

- The rapid increase over the last decade in data volumes, cloud storage, rental computing power, and ubiquitous network connectivity has enabled analysis of operational data that was previously impossible. This transformation is especially seen in new wide-area networks with low power utilization. Industries find themselves facing the prospect of having to use new data in their manufacturing operations.

- The advancement of analytics capabilities. Product development requires analysis for it to be successful, and the stronger and more solid the analysis is, the higher would be the quality of the end product. A lot of analysis is also required to improve efficiencies of business operations.

- The introduction of new form of human and machine interactions. These include the development of augmented-reality systems, and systems that make full use of touch interfaces and other hands-free operating systems.

- The innovations in easing the transfer of digital data to something physically usable. Examples include the improvements in advanced robotics and the onset of 3D printing technology as well as rapid prototyping.

With these drivers at work, industries are finding it increasingly imperative to keep up with the times, especially if they plan to remain competitive.

Four Main Characteristics of Industry 4.0

Proponents of Industry 4.0 name four main and distinct characteristics:

1. *Vertical integration of smart production systems*

    Smart factories, which are essentially the core of Industry 4.0, cannot work on a standalone basis. There is a need for the networking of smart factories, smart products, and other smart production systems. The essence of vertical networking stems from the use of cyber-physical production systems (CPPSs), which lets factories and manufacturing plants react quickly and appropriately to variables, such as demand levels, stock levels, machine defects, and unforeseen delays.

Similarly, networking and integration also involve the smart logistics and marketing services of an organization, as well as its smart services, since production is customized in such a way that it is individualized and targeted specifically to customers.

2. *Horizontal integration through global value chain networks*

   Integration will facilitate the establishment and maintenance of networks that create and add value. The first relationship that comes to mind when we talk of horizontal integration is the one between the business partners and the customers. However, it could also mean the integration of new business models across countries and even across continents, making for a global network.

3. *Through-engineering across the entire value chain*

   The whole value chain in industry is subjected to what is termed through-engineering, where the complete lifecycle of the product is traced from production to retirement. Under other manufacturing disciplines, for example, clothing, the focus would be on the manufacturing process alone, to make the product, sell the product, then ship it and forget about it. There is little concern for what happens to a poorly manufactured shirt for example, let alone what happens to it future sales trends, after the customer throws it in the trash. However, when dealing with industrial components, quality is king. Consequently, there must be focus on quality and customer satisfaction so the manufacturer must build products to meet the customer's expectations. For example, an owner of a Mercedes Benz will expect components manufactured to the highest quality and have after-service support. Industry 4.0 covers both the production process and the entire lifecycle of the product.

4. *Acceleration of manufacturing*

   Business operations, particularly those involved in manufacturing, make use of many technologies, most are not innovative or expensive, and most of them already exist. As can be seen from these four characteristics of Industry 4.0, there is a heavy focus on this concept of a value chain, but what is a value chain?

The Value Chain

All companies strive to optimize their value chain, regardless of size, as they need partners perhaps in design and development, marketing, or with supply chain. The manufacturer's goal is, like in enterprise, to focus on the core disciplines that produce profit and outsource supply, logistics, marketing, and sales. Their perspective is to reduce expenses while maximizing profits. After all, large companies become profitable and successful because they do something better than their competitors, but sometimes even a small company does one task better, as they are focused on a specific task and can do it more efficiently. Therefore, value chain requires that large manufacturers team with partners that have skills in certain disciplines in order to reduce costs.

An example of this scenario is an oil and gas exploration company, which may well be cash rich but doesn't want to invest and learn how to design and manufacture pipes. Instead, it buys pipes from a specialist oil pipe manufacturer, a company that's a fraction of their size and wealth.

However, that is not as easy as it may initially seem. In manufacturing, becoming and sustaining profitable relates to buying raw material at optimum cost to transform that material into a saleable product. Every week competitors are changing their products by price, quality, or availability. Consequently, the goal of large businesses is to identify the core business activities, those that generate most profitable activities of a business to ensure maximum profit, and then outsource the rest.

Therefore, a value chain, by best manufacturing and industrial practice, is mandatory for any producer of goods or services. There are few industrial sized companies, if any, that can support their own value chain, no matter their wealth. Take for example Shell. They require vast amounts of material, whether that is pipes, skilled people, oil tools, oil rigs, helicopters, and even buildings, that they cannot possibly manufacture themselves.

There are two components of a business value chain—horizontal activities and vertical support activities. Horizontal activities directly relate to the manufacturing chain, which relates to each step of the manufacturing process of the product. The vertical support activities, such as IT, sales, and marketing, relate to the production through to the after-sales service.

Primary or horizontal values directly relate to the value that can be added to the production, sales, support, and maintenance of the product. Primary activities add value to the product, and they include functions such as:

- Inbound logistics—These are costs and endeavors that bring the raw materials into the company. They could entail raw material costs, landing costs, taxes, and the cost of storage and distribution.

- Operations—This is where value is added to the raw material by transforming it into a saleable product and this is where profit is determined.
- Outbound logistics—These are the inherent costs associated with shipping products to customers. Therefore, all the functions of storing, distributing, and maintaining stock come under outbound logistics. Traditionally, outbound logistic costs can be very high, hence the initiative to reduce storage and the risk of holding too much stock by the move to produce on demand. Producing on demand or upon a firm order for high cost goods greatly reduces this risk.
- Marketing and sales—Value is also added at the marketing and sales stage, whereby the product goes through the advertising role to convince customers it is a product they desire.
- Service—After-sales the service function considers the value of maintaining a product through its lifecycle.

The other value chain component that must not be overlooked is the support function, and it comprises these reputational features:

- Company infrastructure—This relates to how stable the company is and how reputable are its products, the quality of goods, and their serviceability.
- Human resources—HR relates to how the company manages their workforce. Reputation is built on many factors such as how a company treats their employees. This is a major factor that should never be overlooked. For example, if the company gets a reputation for hiring and firing, the word will soon get around.
- Technology development—This factor relates to the innovation and quality of the technology and engineering teams and their subsequent reputation for producing good, fit for purposeful products.
- Procurement—This is the ability to source and access at reasonable costs a reliable source of raw material or component parts, and this requires good vendor reputations within the business.

Creating a Value Chain

To create a viable value chain, you need a strategy that considers both primary and support activities. The analysis should consider every step of the business process and ensure that it is fully aligned to the company strategy. Who should perform the value chain audit is debatable; some favor high ranking executives with deep company strategy knowledge, others tend to align with a wide mixture of subject matter experts who have proven expertise in their own domain, as this might provide greater depth and width to the audit.

When performing a value-chain audit, it is typical to trace a product from the moment the company procures its components as raw material, through to the production stages, until its eventual sale to the customer. During the audit, the team tasked with the audit should note, during each stage, if there were any possible ways that the process could have been better.

However, the first stage of a value added audit is only the first step in a value added strategy; we must follow up on the initial findings, as there must be analysis and implementation of findings. The result is often a choice between a differential or cost perspective, which is dependent on the company's initial objectives—to improve the product or to cut costs.

Differential Prospective

Manufacturers that want to differentiate their products based on quality, features, or reliability adopt the differential perspective. It may cost more in the manufacturing process, but the company will be relying on customer satisfaction and good customer experience to bring them back for up-sales and cross-sales. Differential perspective entails providing features or quality that supersedes the competition's best efforts and although it might cost more from a production viewpoint, profits will increase due to enhanced reputation and elevated customer experience.

Cost Differential

The alternative to differentiation by quality is to cut costs and product quality. This happens via a strategy to cut production costs and thereby customer prices. However, it is not always the easy option, because a cost audit requires a far more diligent study. Consider one scenario to cut costs rather than improve quality.

Cost-cutting exercises require every stage of the value chain to be audited for efficiency and cost. This entails evaluating every method, process, and procedure to determine the most efficient cost effective means to an end. For example, every stage of the primary and support prospective will need to be audited, to realize the cost burdens, inefficiencies, and potential cost savings. Secondly, there will have to be detailed evaluation studies to determine the realistic cost savings of alternative methods. Thirdly, there is a need to evaluate the costs in changing a process, retraining staff, laying off staff, or implementing new technologies.

However, in its favor, a cost-cutting audit can reveal operational and financial inefficiencies and design/management/production flaws that lead to higher customer costs. Most importantly, those participating in the audit will gain a deep understanding of the company's processes and procedures, which can lead to further insight, a better understanding of the holistic process, and more innovative alternatives.

So why have manufacturing companies not done this before?

Some forward-looking companies partnered with third-party entities in the supply chain through VPNs and Extranets; however, the connectivity was limited to human interaction. With the Industry 4.0 model, cyber-physical systems will interact with one another as well as humans to control the value chain.

Benefits to Business

One of the common misconceptions regarding Industry 4.0 is that it will benefit only the manufacturing industries. However, that isn't strictly true. Yes, manufacturing is the focus, but Industry 4.0's impact is more far-reaching than the boundaries of manufacturing.

Industry 4.0 affects not only the local cyber-physical systems and local industrial processes but the entire value chain, including the producers and manufacturers, suppliers, and workers. One of the initial concerns raised in early adopters of Industry 4.0 is the lack of skilled workers. The education sector will have to step up to produce more talent equipped with the skill sets and competencies required in Industry 4.0. Software and technology developers will also have to look into adjusting their skillsets and becoming more aware of the intricacies of industrial control systems.

Governments, on the other hand, are also doing their share, particularly as they are one of the main drivers in a bid to increase industrial productivity. However, that costs money and huge investment, so governments will have to help industry fund Industry 4.0 initiatives if they expect to reap early benefits. (We will look to EU expectations on ROI later.) Also, when it comes to the infrastructure required for systems to operate successfully and smoothly, for example in the integration of inter-company communications and interfaces, serious funding may be required.

However, we should not fall into the trap of thinking Industry 4.0 is all about vastly expensive robots and CPS that is far from the truth indeed most benefits are achievable from small smart infrastructures, which include those that involve smart mobility and smart logistics.

What are the benefits Industry 4.0 promises for SME? Here are many:

- *Increased competitiveness of businesses:* It can provide a level playing field through cooperation and a confederation of firms. Industry 4.0 is expected to enhance global competitiveness and present a level playing field, with regard to labor costs, but it also enables small companies to work together to challenge large companies. If for example we can reduce the wage bill, it will probably be no longer cost effective to outsource to foreign labor markets for manufacturing and processing. Indeed, experts believe that in ten years our products will no longer be built by a Chinese or Indian worker, but rather by a US/European programmer.

- *Increased productivity:* With the increase in efficiencies, lowering of operational costs that will lead to increased profits. This will also drive forward improvements in productivity levels. Feasibility studies conducted in Europe are forecasting vast productivity gains in de-industrialized nations such as France and the UK.

- *Increased revenue:* The manufacturing sector will reap the benefit of an increase in its revenues. Industry 4.0 is one of the major drivers for the growth of revenue levels and government value-added GDP, even though its implementation will also require significant investment. However, return on investment is predicted as being extremely high, sometimes too optimistic (such as UK manufacturing expected to increase by 20% by 2020).

- *Increased employment opportunities, enhanced human and IT resources management:* Employment rates will also increase as the demand for talent and workers, particularly in the fields of engineering, data scientists, and mechanical technical work, will increase. However, it has to be realized that there is likely to be only a small net gain if any, as traditional labor workers will be either retrained or let go, and not every production line worker is capable of becoming a proficient data scientist overnight. Employment losses will not be restricted to just manual workers. Anyone whose job can be more efficiently handled by an IT service, for example highly paid network and system engineers, will likely be replaced by augmented reality troubleshooting and maintenance systems. However, on the plus side, employment opportunities will not be limited to programmers and data scientists; there will always be work for the industrial process analyst and for supervisors to watch over the integrity of the product lines.

- *Optimization of manufacturing processes:* Integrated IT systems with OT systems is always problematic but within the production process, merging the systems will certainly make the most out of the resources at hand. Administrators can control and streamline processes, and this will enable collaboration between and among producers, suppliers, and other stakeholders along the value chain. The usual time that it takes to produce one unit will decrease, making the process more efficient since the steps required are simplified, without compromising quality. Decision-making is also done in real-time, which is imperative in industrial scenarios. Similarly, those vertical IT elements come into play as business segments are allowed to develop their full potential as they are influential. Each technical CPS in the context of a system rather than a single machine, has its own point of view of the holistic process yet they can understand the needs of their customers or the partners with whom they collaborate.

- *Development of exponential technologies:* Industry 4.0 will provide a platform for the basis of further innovation with developing technologies. Suppliers and developers of manufacturing systems and technologies will use them as basis on what to develop next. We have seen this with mobile phone applications as, for example, more developers are using open APIs to mash up applications. Already, developers are looking into technologies that will be an improvement on the current, GPS, RFID, NFC, and even accelerometer sensors embedded in the standard smartphone.

- *Delivery of better customer service:* Industry 4.0 monitoring and feedback mechanisms rely on the industrial concepts and methods of real time. These concepts applied to logistics and dashboard reporting is very important as they generate and analyze in close to real time. Therefore, dashboards of key business indicators are available immediately, which allows decision makers to realize the current state of affairs and make intelligent decisions and respond faster to the industrial process and ultimately to the needs of the customer.

Industry 4.0 Design Principles

One of the basic principles of Industry 4.0 is connecting systems, machines, and work units in order to create intelligent networks along the value chain that can work separately and control each other autonomously but in a cohesive manner.

Industry 4.0 has six identified design principles that manufacturers and producers use in their automation or digitization efforts for their production processes. They are discussed next.

Interoperability

The production process does not simply follow a predetermined set of methods or steps and involve only the people, machines, and processes that are directly involved. Interoperability requires an entire environment with fluid interaction and flexible collaboration between all the components. For example, assembly stations are not separate from the products created or the people who are working on them. Interoperability refers to the capability of all components to connect, communicate, and operate together via the Internet of Things. This includes the humans, the smart factories, and the relevant technologies.

Virtualization

The monitoring of the actual processes and machinery takes place in the physical world and returning sensor data, will then be linked to virtual models or models created via simulation. Process engineers and designers can then customize, alter, and test changes or upgrades in complete isolation, without affecting the physical processes they have virtualized. Producers in the setup of Industry 4.0 will use the creation of a "virtual twin" of the smart factory to greatly enhance existing processes and products and reduce product development and modeling, creating a production process and hence reducing the time to profit of new products.

Decentralization

Industry 4.0 supports decentralization, which enables the different systems within the smart factory to make decisions autonomously, without deviating from the path toward the single, ultimate organizational goal.

Real-Time Capability

Industry 4.0 efforts are also centered on making everything real time, which requires that the production process the collecting of data and the feedback and monitoring of processes is also achieved in real time.

Service Orientation

The Internet of Things creates potential services that others can consume. Therefore internal and external services are still going to be required by the smart factories, which is why the Internet of Services is such an important component of Industry 4.0.

Modularity

Flexibility is also another design principle of Industry 4.0, so that smart factories can easily adapt to changing circumstances and requirements. By designing and building products, production systems, and even the conveyor belts that are modular and agile, the smart factory is flexible and can change in production. Producers can ensure that individual product lines can be replaced, expanded, or improved on with the minimum of disruption to other products or production processes.

Building Blocks of Industry 4.0

There are nine identified technological trends that are said to be primarily instrumental in shaping industrial production. The following sections discuss each.

Big Data and Analytics

These days, the manufacturing sector is finding itself inundated with an increasing amount of data from various sources, and there is a need to gather all that data, collate and organize it in a coherent manner, and use the analytics provided by the data sets to support management's decision-making. Business cannot afford to ignore the data coming in, as they might prove to be very useful when it comes to optimization of production quality and service, reduce energy consumption, and improve efficiencies in the production process.

For example, data can be collected from the various phases of the production process. These large amounts of data will be analyzed in correlation with each other in order to identify phases with redundant processes that may be streamlined.

To sum it up, there are six Cs in big data and analytics with respect to the Industry 4.0 environment. They are:
- Connection, which pertains to sensors and networks
- Cloud computing
- Cyber, which involves model and memory
- Content/context
- Community, or sharing and collaboration between and among stakeholders
- Customization

Autonomous Robots

The use of robots in the manufacturing process is no longer new; however, robots are also subject to improvements and evolution. Creators of these robots are designing them to be self-sufficient, autonomous, and interactive, so that they are no longer simply tools used by humans, but they are already integral work units that function alongside humans.

Simulation

Previously, if manufacturers wanted to test if a process was working efficiently and effectively, trial and error was required. Industry 4.0 uses virtualization to create digital twins that are used for simulation modeling and testing and they will play more major roles in the optimization of production, as well as product quality.

Horizontal and Vertical System Integration

Having fully integrated OT and IT systems is something that Industry 4.0 aims for. The goal is to create a scenario where engineering, production, marketing, and after-sales are closely linked. Similarly, companies in the supply chain will also be more integrated, giving rise to data integration networks, collaboration at automation levels, and value chains that are fully automated.

The Industrial Internet of Things (IoT)

Embedded computing and networking will connect transducers and devices and these are an essential part of Industry 4.0. The industrial Internet of Things will make this possible, since transducers and field devices designed for the IoT and equipped with low-power radio networking to enable them to interact and communicate with each other, while also becoming connected with a gateway to a control and management layer, will become ubiquitous throughout the Smart Factory and supply chain.

Cyber-Security

Industrial systems are becoming increasingly vulnerable to threats, as can be seen by recent attacks on industrial targets in 2015. To address this, cyber-security measures have to be put in place that recognize the new vulnerabilities and challenges that integrating industrial control processes and systems with the Internet produces.

The Cloud

The large data sets involved in Industry 4.0 means data sharing will be not only desirable but imperative to leverage the full possibilities within the value chain. However, few manufacturing plants will have the storage capacity to store and analyze the vast amounts of data collected. Fortunately, cloud service providers do have the capacity and can create private clouds suitable for manufacturing data storage and processing.

Additive Manufacturing

Additive manufacturing such as 3D printing enables manufacturers to come up with prototypes and proof of concept designs, which greatly reduces design time and effort. Additive manufacturing also enables production of small batches of customized products that offer more value to customers or end users, while reducing cost and time inefficiencies for the manufacturer.

Augmented Reality

Businesses are increasingly looking to reduce the maintenance and training overheads associated with production, marketing, and after-sales support. Manufacturers are turning to augmented-reality-based systems to enhance their maintenance procedures while lowering the costs of having experts onsite.

Industry 4.0 Reference Architecture

As the Industry 4.0 environment is not envisaged as being typically just one big company but a collection of SMEs, all with their unique disciplines forming supply chain partnerships, the reference architecture must reflect this. What this means is that the architectural model is that it should be a general model that each company can deploy to enable one common approach to data protocols and structures, machine, devices, and communication interfaces.

The difficulty is that each company in the partnership has a different perception, and these can be from a manufacturing, software, engineering, or network point of view. For example, from a manufacturing perspective, the focus will be on process and transport functions. Whereas an IT software perspective would consider the type of application and system management software each partner uses. There interfaces are required between ERP, business management, software applications to enable inter-company logistics and business management planning. From the network side, the perspective is on devices and how they connect. The list of devices that need taken into consideration is vast, as they can be sensors, actuators, servers, routers, PLCs, field bus, Profinet, tablets, or laptops. The engineer perspective is more about the product lifestyle management. Consequently, the use of existing standards is a requirement within the Industry 4.0 architecture.

The use of existing standards is important for connectivity across the entire value chain and for allowing companies to inter-connect and participate in an autonomous or semi-autonomous way, which is part of the value of Industry 4.0.

Smart Manufacturing

Lean manufacturing, which has been the popular process improvement strategy over the last decade, was aimed at producing initiatives to eliminate overburden, inconsistencies, and waste. The goal of lean manufacturing was to produce goods smoothly and consistently. Smart manufacturing is similar as it too is a process improvement initiative; however, its aims are to merge the digital and analogue worlds by building connectivity and orchestration to provide an enhanced processes that delivers goods smoothly and consistently.

Therefore, lean and smart are complementary process improvement strategies that we will see working side-by-side in the factories of the future.

However, before we can realize the collaboration of lean and smart manufacturing, we need to address some disconnects within smart manufacturing and understand how it will be deployed in the workplace.

Equipment

The first issue with smart manufacturing is going to be connecting all the machinery and equipment. CPS and advanced sensors work together and connect to systems that can control and orchestrate the production process by sensing the condition of the product being manufactured. To reap benefit from this connectivity will require collecting and processing real-time data from the machinery, CPS, and sensors in order to analyze and determine the status of each product. Yet many companies still do not have sufficient technical and decision-making structures in place to benefit from the connected processes and the data they produce. An example of this type of failure is when companies fail to analyse the data available to predict the maintenance of machinery and strategically plan servicing so that it reduces the effective downtime of a production line.

Redefine the Workforce

People are as equally important to smart manufacturing as machinery, as they connect the customers to the manufacturing process as well as ensure the production line is running smoothly. As a result, technicians, inspectors, and process operators must be connected via function-specific applications. Managers also need connectivity to the manufacturing process via dynamic real-time dashboards that display status, trends, and alerts. And don't think that because all those smart machines and systems are automating most of the mundane decision making, that there will be no need for people. They also trigger action-based exceptions and anomalies that require expert analysis.

Consequently the workforce of the future will most likely require multi-disciplined generalists with computer, mechanical, and process engineering skills. This is because these workers will have to operate across multidisciplinary walls, connect smart factories together, retrieve and interpret data from analytics, and even protect supply chains from security threats and intrusions.

There are many manual steps required before you can create a digital thread that has the links and translations between the 3D design process and the configuration of the physical manufacturing machinery and inspection processes. For example, if we consider 3D printing the digital thread in this instance relates to the links between the product design and the actual 3D printing process.

The steps might include simulation of the virtual product in collaboration with partners and customers before anything is actually configured or supply chain specifications produced or revised. Creating 3D simulations and proof of concept models can save a lot of time and money by ensuring the product is exactly as it should be before the manufacturing processes get underway. Another benefit from the digital thread is that actual manufacturing processes can be simulated to make sure that they will run safely and efficiently, removing any doubt or trial and error from the process.

Products

One of the goals of smart manufacturing is that parts and components should be identifiable and smart; for example they should be able to identify themselves and hold important information. By using RFID tags that are embedded within the product or packaging, smart machines will be able to identify the component and read its status information and recognize the configuration of each product as it traverses the production line.

By doing so, machines can load the correct parts and programs to handle each component dependent on their configuration. An example of this is a production line, filling shampoo bottles. Each bottle may be a different version of a brand and requires different coloring or perfume added at a specific stage. If the machine can read an embedded RFID tag on the bottle that identifies its brand variety, the machine can automatically add the correct color and perfume for that particular product. That is a very simplistic example, but the point is if every part on the production line carries its identifier and a record of its own history then production decisions can be made intelligently during the manufacturing process. This enables customization of products during manufacturing, which can lead to profitable production of even lot sizes of one.

Of course, there is a slight issue here, as business processes require a flow of tasks with the output from one task being the input of the next task. Traditionally, on a production line that would be relatively straightforward and a paper exercise. However, smart manufacturing and smart parts and products can influence the actions of the production line, which means the business processes become more complex.

Business Processes

Industrial Internet business processes will need to cater to parallel processing, real-time responses to process control, and collaborative machines (CPS) on the production line. This leads to automation of processes, which is ideal as that reduces overhead related to maintaining multiple ways to achieve the same result, for example separate lines for each version of the shampoo in the earlier example.

Optimizing the Supply Chain

The traditional paper exercise of managing supply chains is no longer efficient enough to support the smart manufacturing process. Therefore, instead of buyers working away phoning, e-mailing, raising documents, and faxing orders, the process must become automated and digitalized. Suppliers not only need to connect via business-to-business communication channels, but must develop a demand-driven model that reduces inventory and replenishes material based on sensing parts stock movement and counts.

Manufacturers are under more pressure from customers as the trends shift from make-to-stock toward the more efficient make-to-order, configure-to-order, or even engineer-to-order. These new trends in logistics and stock management place a strain on traditional production lines. Previously production lines were set up and configured to do large runs, for example, 500,000 shirts, and it was economical and practical to devote an entire conveyor line to that customer order.

However, when the customer shifts to build-to-order, as Dell Computers and Cisco did in the early 2000s, this creates a problem. For Dell Computers there was probably enough consumer orders to keep their manufacturer's production line busy, but with some of Cisco's highly specialized and expensive equipment, it would require a stop-start production line. This was because production lines needed to be configured and set up for different products and they are not agile and flexible to change at short notice. This is where smart machine provide a real benefit, as they can be reconfigure dynamically by simply uploading the process software, templates, and parts.

The orchestration of multi-tier supply chains is not a trivial task and requires complex processes that are connected to design product documentation and, very importantly, to the change management process. On many occasions, engineers, designers, and process engineers have (when working in collaboration with partners and customers) agreed on a change control function that altered the component specifications without instructing the supply chain, with calamitous results.

Customers

In the Industrial Internet, the customers' evaluation of the product is everything, it is termed the customer experience. Whether that be through product evaluation or simply the experience of using the product, the goal is for the customer to gain value and a satisfying product and for the producer to gain a good reputation.

However, customer experience is very difficult to evaluate. Yes we can have surveys and ask the questions that relate to a product. However, we have seen that doesn't always work. A prime example is that of a global distributor of cola fizzy drinks. This company decided that their competition was proving to be aggressive and their advertising campaigns were troubling their traditional markets. As a result they changed their formula, in order to revitalize their brand. Unfortunately, they failed to understand was that changing the recipe also changed not only the taste but, more importantly, the brand. The result was a huge shift from their traditional product to their competitors.

Now, why did this come about it? After all it makes no sense whatsoever. However it appears that the incentive was based on customer experience. The cola company did massive research into taste surveys of the public that deemed the new product better tasting, yet it still bombed when it was released.

This is why big data and in-depth analytics is so important; it is no longer about a CEO or her team determining the future of a product; it is about deep mining research and sales figures which do result in true figures, for sales and customer experience.

Big Data analysis is hugely important. It can produce information that is business specific and produce knowledge that is vital to the product's design and the customer's experience.

Why is customer experience important? Well, it is why you do business, and it is due to a positive customer experience that you make profit.

Customer Experience

Customer experience relates to the individual's acceptance that the product they bought is value for money. The acceptance is based on the quality of the product, the fitness for use, and the cost in relation to competitive products. However, customer experience also relates to the way the customer relates to the goods, the seller, and how they feel and trust about the relationship between the parties. For example, used car salesmen have atrocious reputations. However, this doesn't mean that every car they sell is a dud—most actually are good deals because if they weren't, customers would not come back.

Say you buy a smartphone that costs $500, and then you discover that every application that you need has to be bought through the manufacturer. Would you be happy? The result would be a poor customer experience.

So how do we manufacture only products that our customers want?

Through smart manufacturing and clever design; hence the need for smart factories.

References

http://www2.deloitte.com/content/dam/Deloitte/ch/Documents/manufacturing/ch-en-manufacturing-industry

www.siemens.com/digitalization/

www.gtai.de/.../Industries/industrie4.0-smart-manufacturing-for-the-futu..

http://motherboard.vice.com/read/life-after-the-fourth-industrial-revolution

CHAPTER 14

Smart Factories

The heart of Industry 4.0 in conceptual terms is the Smart Factory (Figure 14-1) and everything revolves around this central entity that makes up the business model. If we look at how Industry 4.0 will work in theory, we can see that everything from the supply chain, business models, and processes are there to provide the Smart Factory. Similarly, all the external interfaces from supply chain partners, smart grids, and even social media conceptually have the smart factory at the hub—it is the sun around which other processes orbit.

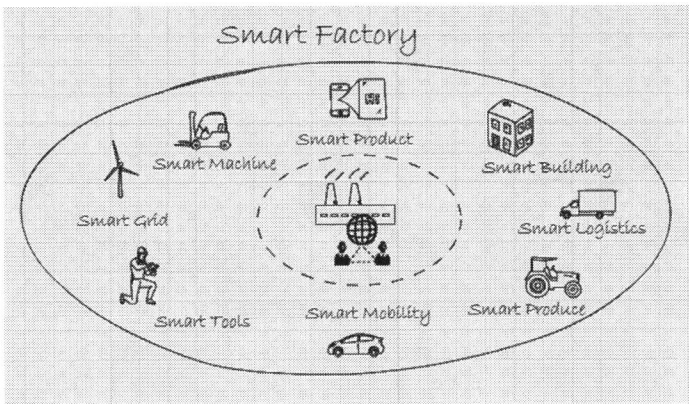

Figure 14-1. Smart Factory

So what is a Smart Factory and why is it so important to the future of manufacturing?

© Alasdair Gilchrist 2016
A. Gilchrist, *Industry 4.0*, DOI 10.1007/978-1-4842-2047-4_14

Introducing the Smart Factory

A Smart Factory hosts smart manufacturing processes, which we have explained previously. A Smart Factory is futuristic in that it can produce and deliver productivity well beyond our expectations. If we look at how this is possible, we can see that Smart Factories are a bringing together of technologies that provide the optimum methods and techniques in manufacturing. Furthermore, we can witness that Smart Factories are not just intelligent machines and robots communicating through an advanced software product. Indeed, these machines have advanced beyond M2M and are not just collaborating but also communicating through advanced software, algorithms, and industrial processes. However, it is important to realize that Smart Factories, like Smart Homes, are not some futuristic vision—they are with us today and have been for a decade at least.

Therefore, before we go any further, let's look at how a Smart Factory works, and then we can see perhaps the benefits and the massive improvements in efficiency and productivity that they can bring.

As an example, let's take a production line that different lines of shampoo. In this production line scenario, a smart machine fills each bottle with the same base ingredients. Each variant of the brand may have different color additives or perfume added to align itself with the products intended market.

In traditional manufacturing, this would require a production line for each individual product. The production process would specify that each bottle be sent down a production line to a dispenser machine, which would fill the bottle with the required mixture of ingredients as required. However, if we have many minor varieties of the shampoo brand then this is hugely inefficient, as many machines are doing the same work. Why can we not identify each variety of the product coming along the line and fill it as required?

This is one of the basis of Smart Manufacturing, because we can reduce waste and inefficiency by identifying products on the production line and determining their status and what is more their history and what specific stage of production they must next pass through.

Now how do we do this? We can use RFID tags that are so miniaturized that now they can be embedded into a label or use NFC (near frequency contact) such as in card payment systems. NFC is a bit fragile it requires close proximity to the reader, whereas RFID is astonishingly capable. Take for instance a racing car embedded with an RFID tag and during each lap, an RFID reader counts the number of laps. Incredibly, an RFID reader can count reliably every lap a racing car performs even at speeds of 200 mph and more. Therefore, RFID tags are perfect for Smart Factory applications where the speed of the production process must not be compromised.

So let us see how a Smart Factory can work in practice as suggested by DR. Uwe Dittes of SAP SE in his lecture on the subject of Technical and Operational Solutions for Industry 4.0 in ERP.

Smart Factories in Action

In this scenario, we will consider the production of shampoo variants, and how we can construct a production line to produce all products even though they differ in label, color, and perfume. The problem is of course is that the machinery must be able to identify and classify each product traversing the production line. Only by doing so can they possible decide on a course of action suitable for each entity. The way to do this is through individual identification and data storage via RFID tags on the products themselves. In this way each product, knows what it is, how old it is, and what the next stage of manufacturing should entail. It might also carry much more information such as the idea storage conditions or handling methods, which is advantageous to not just the manufacturing process but the entire product lifecycle.

Subsequently, let's see how this works in practice on a smart manufacturing line.

To begin, consider how those different bottles of shampoo would have been produced on a traditional Industry 3.0 production line. This would have required three resources, one controller, and a supervisory system. Figure 14-2 shows the three levels of the production line. At the lowest physical layer are the production resources, above them is the Manufacturing Execution System (MES), and one layer above that the Enterprise Resource Planning system.

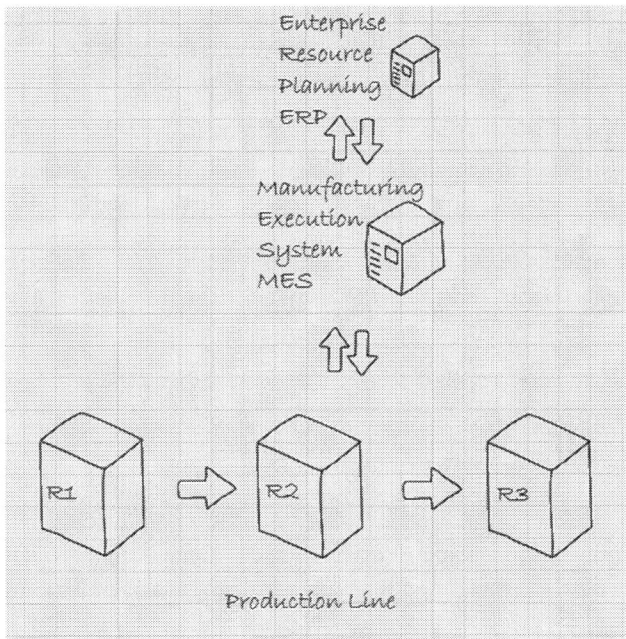

Figure 14-2. Production line diagram

Chapter 14 | Smart Factories

From Figure 14-2, we can see the three production resources required to make our shampoo products. The first resource R1 makes and stores the base ingredients. The second resource R2 receives a controlled amount of the base liquid, which it mixes with variant specific color additives, perfume, and chemicals/nutrients. Resource R3 receives the mixture from R2 and fills the appropriate bottle.

The ERP system controls production level by monitoring the sales orders generate by the reseller chains and supermarkets and sends instructions to the MES to manufacture the appropriate amounts to fulfill the orders. The MES initiates production to fulfill the orders and provides feedback of the production status to the ERP system.

Production works this way in a standard modern factory. However, it is not without its flaws.

Consider some of the weaknesses. The first weakness lies in the serial production line in so far as if one resource fails, then the whole production line fails. Secondly, any failure in the ERP or MES will also block production. Expansion or reconfiguration of the production line is difficult due to the interfacing difficulties between the MES and resources, as there can be hundreds of interfacing options. Similarly, interfacing with the ERP system can be complex due to their monolithic architecture. Thirdly, although it is highly desirable, it is not always feasible to have the ERP system updated in real time by the MES on the status of production; for example, the number of bottles produced and the number still to run to fulfill an order.

Industry 4.0 can mitigate some if not all of these weaknesses. The way Industry 4.0 works is that the resources in the previous example are replaced by CPSs, as shown in Figure 14-3.

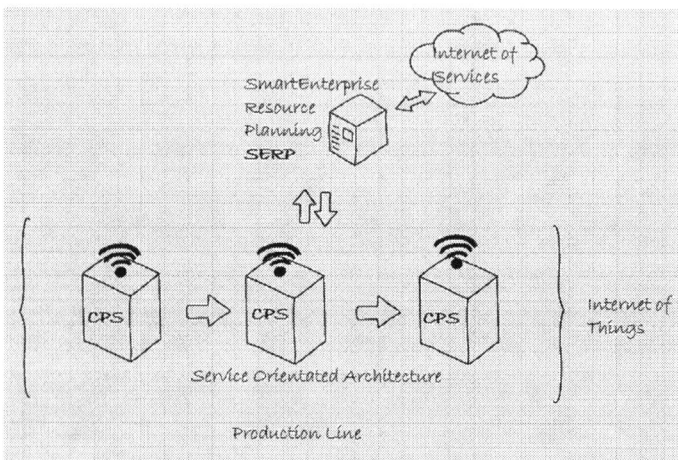

Figure 14-3. Revised production line with CPS

What happens here is that by replacing resources with CPSs, the strict serial line is no longer fixed; it becomes flexible, as CPSs are intelligent and responsive. CPSs have embedded sensors and can communicate with one another over wireless radio links, which enables one CPS to take over the tasks of a failed CPS. This ability for CPS to self-diagnose and check the status of the production line and then take appropriate collaborative action provides for improved availability and resilience. Additionally, because CPS interfaces with one another directly, they do not require a MES system so that removes another potential point of failure. More importantly, removing the MES mitigates the issues with interface mismatching and reconfiguration, which was a major pain point with the previous topology.

The self-sufficient CPS are not the only smart, intelligent entities on the production line, the product also is smart. For example, the shampoo bottles will be fitted with RFID tags, which identify which brand and variant it is, and the state of its production (its own production history to date), as well as the next stage it has to proceed to in order to complete its production. Moreover, the product's intelligence extends beyond the production line, even the smart factory, into the warehouse and later into the reseller chain. Furthermore, the product's intelligence stays active even in the service of the customer. For example, consider a far more intelligent product than the humble shampoo bottle, such as a tractor engine, the smart engine could during its productive working life self-diagnose and alert the customer or service department of its maintenance status and even predict failure of a component. This is an important shift from break-fix to fix-before-break and can greatly improve service availability and downtime.

With the MES becoming redundant to our smart design, the ERP system now becomes a smart ERP (SERP), and it communicates directly with the CPSs to control the production of products to fulfill the order books. By being directly connected, the SERP system now learns in real time the status of production, CPS health, and other real-time sensor data. The SERP accomplishes this by using in-memory databases for real-time streaming analytics to enable business processes to run faster and better.

Thus the concept of the smart factory was born; it's a system that is flexible, agile, and intelligent that reaches beyond the CPSs and the walls of the factory and into the products, and hence to the entire value chain.

Why Smart Manufacturing Is Important

Creating this manufacturing revolution requires significant collaboration among companies, governments, and academic institutions. For example, in the EU and the United States, they have set up initiatives to fund and encourage smart manufacturing. The EU in particular is striving to re-industrialize and create a level of parity across a very diverse manufacturing capability of member states. Germany and Italy are modern industrial powerhouses that have well developed Industry 4.0 programs. Britain and France, on the other hand, have been de-industrializing for the last three decades and require a massive effort to re-industrialize. Ironically, it is France and the UK that are most likely to benefit from smart factories, as they can bring their manufacturing back onshore and subsequently enjoy great savings in costs and efficiency. In fact, Germany is unlikely to contribute much to the EU targets for increased efficiency and value-add to GDP, as they are already near optimum efficiency levels. The UK and France, however, can make significant contributions as their current performance levels in manufacturing are underperforming and therefore these countries' manufacturing efficiency are ripe for improvement.

In a global initiative, the industrial internet consortium is sponsoring a number of pioneering collaborative projects, called testbeds, which focus on different steps of the manufacturing process. For example, Infosys, working with Bosh, PTC, and Intel, are collaborating on an effort called the asset efficiency testbed. The term *asset efficiency* refers to reducing waste and improving the maintenance and uptime of any industrial asset in operations, maintenance, service, information, and energy. The testbed project is focusing on ways to use data from equipment and processes to give aircraft landing gear maintenance engineer's information with which they can forecast and correct potential failures.

Winners and Losers?

With the adoption of Industry 4.0 or the Industrial Internet and the consequent move to smart manufacturing, optimized supply chains and smart factories there will ultimately be winners and losers. Typically, developed countries such as the United States and Western European member states will benefit more from smart manufacturing and smart factory initiatives. This is predominantly going to be through reductions in operational costs, increased efficiency, and increased productivity. Similarly, companies that form part of the smart ecosystems surrounding these manufacturers will also benefit as part of a symbiotic relationship.

Furthermore, manufacturers in high-wage countries that for many years found it cost effective to outsource the manufacturing to China, India, Brazil, Russia, or to Eastern European countries will now be able to take a different approach. Industry 4.0 will make manufacturing in developed countries much more cost effective and will mitigate the low wage advantage of competitors. With wages having a reduced relative importance to the overall operational expenses, there will be a reversal of the outsourcing trends of recent decades. Countries such as the United States, UK, and France can start to reindustrialize and bring home much of the manufacturing that they have been outsourcing abroad.

Developing countries with low wages that embraced the manufacturing chore from the more developed countries will inevitably lose out in the short term. This is why China and India have recognized the importance of the Industry 4.0 revolution and are committed to change through their "Made in China" and "Made in India" initiatives. South American, Asian, and Eastern European nations will find that they too will need to embrace the fourth industrial revolution as their economies, which are built on the back of industry 3.0 practices will no longer be sustainable, when the low wage advantage is removed from the equation.

Real-World Smart Factories

The last section covered an introduction to smart factory, including what it was and a simple production scenario. This section looks at some state-of-the-art smart factories that are already in production.

GE's Brilliant Factory

General Electric's "Brilliant Factory" already uses IIoT technology in some of its manufacturing plants and aims to introduce smart technology ultimately within the company's 400 manufacturing and service plants. GE aims to transform their traditional manufacturing plants by digitally connecting teams designing products to the factory floor, other supply chain partners and, finally, to service operations. Subsequently, the company aims to build a continuous loop of real-time data sharing, enabling faster, more accurate decision-making, which results in an increase in productivity. This digital thread that they want to create will allow GE to design, manufacture and service their products better. Indeed GE is targeting KPI (key performance indicators) of 20% increase in uptime and 10% rise in throughout.

From App to Production

GE is aggressively undertaking dozens of pilots, which are now underway in their manufacturing plants and supply chain; for example, the company is testing various stages of the process along the value chain. As an example, GE recently developed an app for use in the company's turbo-machinery businesses. The mobile app connects the factory floor with engineering, allowing the engineers to verify that the factory can feasibly manufacture the virtual products, which are still in the design phase. How this works is that when a designer creates virtual construction of a part, for instance a CAD drawing, the app will provide real-time feedback, such as whether the thing can be made and, if not, which features should be adjusted. When a pilot is successful, with all the kinks ironed out, GE will roll out the smart system across factories with similar requirements.

Airbus: Smart Tools and Smart Apps

Another example of a real-world smart factory is the aircraft manufacturer Airbus's plant. Airbus is working in with National Instruments to create what they describe as the factory of the future. The first phase of the smart system targets the reduction of mistakes and wasted time and materials, a lean manufacturing concept. Savings are achieved by giving workers real-time "toolbox talk" information on the job they are tasked to complete. This toolbox information is fed online on demand to the worker's mobile device. This ensures that the workers know exactly what is required and that they are using the correct tools, at the right setting and they know exactly what steps they should be taking at any given time. This use of educational technology has many use-cases within industrial maintenance and reduces the costs of having highly skilled, experienced maintenance engineers onsite. Instead, online augmented reality tutorials in collaboration with sensors on smart tools will transmit information via Wi-Fi to workers who are for example, routing a cable loom in an aircraft, letting them know when they are within a few millimeters of a device to be connected and of the precise connections and wires to attach.

In another scenario, which is common in manufacturing, and something that robots are not very good at either, smart tools detect the amount of torque used, say, to tighten bolts. Over-tightening or under-tightening bolts and screws is a common cause of material failure, but with smart tools this should allow workers to apply the correct amount of torque while at the same time allowing engineers to analyze the data, detect whether the task was done correctly and, if necessary, take action.

Siemens' Amberg Electronics Plant (EWA)

In the vision of Industry 4.0, the physical and the virtual manufacturing worlds will merge. Factories will then be largely able to control and optimize themselves, because their products will communicate with one another and with production systems in order to optimize manufacturing processes. Products and machines will communicate and collaborate to determine not just how a product is assembled but also which items on which production lines should be completed first in order to meet delivery deadlines. Independently operating computer programs known as software agents will monitor each step and ensure that production regulations are complied with.

The Industry 4.0 vision of smart factories also foresees the capability to profitably manufacture one-of-a-kind products, as they will produce quality items quickly, and inexpensively, with little to no production line reconfiguration.

Siemens is the world's leading PLC supplier, and the facility, which Siemens established in 1989, produces Simatic programmable logic controls (PLCs).

Industry 4.0

The devices are used to automate machines and equipment in order to save time and money and increase product quality. They control ski lifts and the onboard systems of cruise ships as well as industrial manufacturing processes in sectors from automobile production to pharmaceuticals. Amberg (EWA) is the company's showcase plant for these systems.

Defects: An Endangered Species

EWA has some quite staggering KPI statistics such as production quality is at 99.99885%, and that is an incredibly low level of defects on a production line. Furthermore, a series of test stations detect the few defects that do occur during production. To put these figures into perspective, we have to consider that the EWA factory manufactures 12 million Simatic products per year. At 230 working days per year, this means that the EWA produces one control unit every second, with a defect rate of 0.00115%.

Computers Producing Computers

Of course, in a smart factory the aim is always to ensure that production is largely automated and in Amberg that is the case, with machines and computers handling 75% of the value chain autonomously. The rest of the work is done by people. For example, all of the production line tasks are accomplished by autonomous machines, driven by PLC controllers. It is only at the beginning of the manufacturing process that anything is actually handled by human hands, and this is when an employee places the initial component (a bare circuit board) onto a production line conveyor belt. However, from that point on, everything runs automatically and autonomously. The notable feature here is that Simatic units control the production of Simatic units as this is the Siemens PLC showcase, so the PLCs create other PLCs on the production line. In the EWA, around 1,000 PLCs are in use during production, from the beginning of the manufacturing process to the point of dispatch.

Industry 4.0 in Action

If we take a closer look at the production process, we see that at the beginning conveyor belts take the bare circuit boards to a printer, which uses a photolithographic process to apply a lead-free solder paste. Subsequently, robotic placement heads mount individual components, such as resistors, capacitors, and microchips, onto the circuit boards. In Amberg, the fastest production line can mount 250,000 components per hour, a process that is controlled by Simatic units. Once the placement and soldering of components is completed, the conveyor belt carries the populated printed circuit boards for testing. When the PCBs arrive at the optical test station, a camera examines the position of the electronic components to ensure everything is in the correct place and no components are missing. Simultaneously, an X-ray machine inspects the quality of the soldered connection points to verify the quality of the joints and to identify any dry joints or poor connections. The next step is where each printed circuit board is mounted into housing and then passed to the dispatch stations.

The final stage is to retest each product and then dispatch them to a delivery center in Nuremberg. The PLCs produced in Amberg are shipped to more than 60,000 customers all over the world. The biggest customer is China and they purchase around 20% of the PLCs produced; the rest are mainly sold to customers in Germany, the United States, and Italy.

Humans in Charge

Although production in Amberg is highly automated, people ultimately make the decisions. For example, an electronics technician supervises the test station for populated printed circuit boards, even though he himself does not test the components and circuitry. The technician uses a computer to monitor the entire value chain from the workplace because each circuit board has its own unique barcode that lets it communicate with the production machines. More than 1,000 scanners document all of the manufacturing steps in real time and record product details such as soldering temperature, placement data, and test results. This wealth of product specific data is at hand for every PLC produced at Amberg.

Around 50 million pieces of process information are generated each day and stored in the Simatic IT manufacturing execution system. This enables technicians and inspectors to monitor and observe every product's lifecycle down to the last detail.

In spite of its highly automated processes, EWA nevertheless relies on people for the development and design of products, production planning, and the handling of unexpected incidents. That isn't likely to change in the near future, and despite Amberg being a model for smart factory technology, it is a long way off being a lights-off production center as around 300 people work per shift, and the EWA has a total of about 1,100 employees.

The Amberg Electronics Plant is an advanced example of Siemens Digital Enterprise Platform—where products control their own manufacturing processes. As we have seen with smart products, their product codes tell production machines what requirements they have and which production steps come next. This smart product system marks the first step toward the realization of Industry 4.0.

Employees with Ideas

Siemens though is not aiming to build a workerless factory. The machines themselves might be highly efficient at what they are programmed to do, but they do not come up with ideas for what products should be made or for ways to improve the system. At EWA the employees' suggested improvements account for 40% of annual productivity increases. The remaining 60% is a result of infrastructure investments, such as the purchase of new assembly lines and the innovative improvement of logistics equipment. The important point here is that it appears that employees are much better than management at determining what works or doesn't work in daily operation and how processes can be optimized. In 2013, the EWA adopted 13,000 of these ideas and rewarded employees with payments totaling around €1 million.

Siemens Chengdu

Such has been the success of EWA in Germany that Siemens decided in 2013 to open a new smart production center in China. The Siemens Electronic Works Chengdu (SEWC) in Southwest China opened in February 2013. Many parts of the plant replicate its Amberg counterpart, and Simatic controllers are also made there as China is Siemens biggest market for PLCs. Taking the product closer to the customer also puts the focus on China, which is the world's largest market for automation technology and second only to the European market for programmable logic controllers.

Interestingly, the software tools and production sequences are the same in Chengdu and Amberg. Despite Amberg being now over 25 years old. However, even though the production area has remained unchanged and the number of employees has hardly increased, the plant now manufactures seven times as many units as it did in 1989. More importantly, quality has increased substantially as well. Whereas the production facility had 500 defects per million (dpm) back in 1989, it now has a mere 12 defects per million.

However, SEWC is still under development and currently its level of automation is still lower than Amberg's 75%. Additionally, SEWC has not yet had the capacity to produce a similar range of products. Where SEWC is comparing favorably with the Amberg Electronic Works is where energy efficiency is concerned. The SEWC plant recently received LEED (Leadership in Energy & Environmental Design) certification at the gold level. Compared with similar buildings, the plant uses about 2,500 metric tons less water, discharges about 820 metric tons less CO_2, and saves about €116,000 in energy costs each year thanks to smart building technology. The plant is the first in Chengdu to receive LEED Gold certification.

Industry 4.0: The Way Forward

Manufacturing is a vital component in the economies in both developed and developing nations. However, the industry has and is still going under rapid areas of change, bringing new opportunities and challenges to business leaders and policy makers. See Figure 14-4.

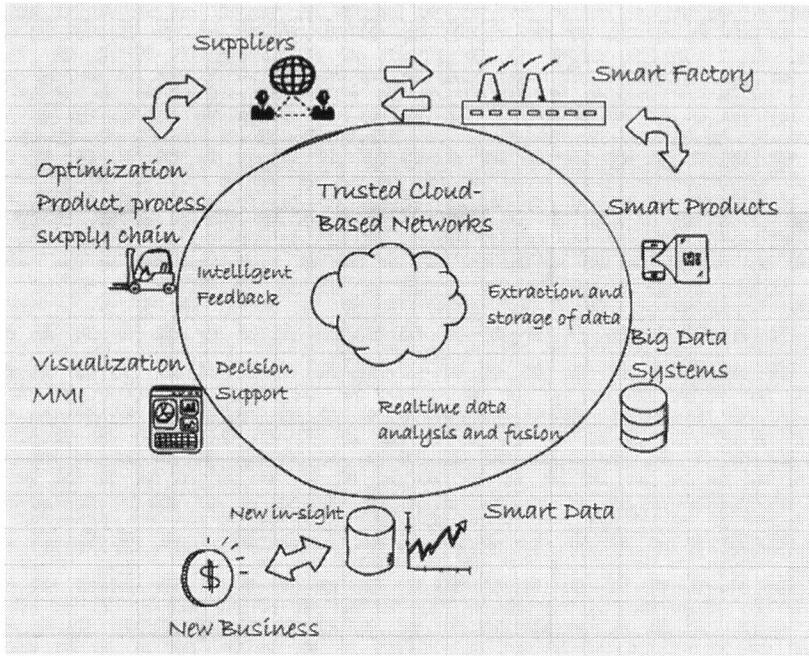

Figure 14-4. New Internet services and business models

It has been a tumultuous decade for manufacturing in both large developing economies, as a result of a severe recession that choked off demand. In advanced economies manufacturing demand, productivity, and profits slumped, which acted as a catalyst for employment falling at an accelerated rate.

Despite the global recession, manufacturing remains critically important to both the developed and developing nations. In the industrialized developing nations, manufacturing provides a significant contribution to the economy as well as a path from a subsistence agriculture culture to rising incomes and living standards. In the developing nations, it remains a vital source of employment, as one job created in manufacturing can lead to two jobs created in the supply chain. Furthermore, in developed industrial nations value-add is circa 20% of the GDP, yet manufacturing contributes disproportionately to a nation's growth as it inspires innovation and competitiveness, and this leads to sponsorship of research and development, which benefits academic and technical institutions. Indeed, in the United States, manufacturing's contribution to GDP is 20% yet it contributes 77% to private sector research and development. Manufacturing also makes an outsized contribution to exports with a 70% share, and productivity growth of 37%.

In developing countries, we can see how important industrialization is to economies as wages rise and productivity grows. However, at around 20-30% of GDP that sharp rise in employment, wages and relational productivity to the GDP plateaus before falling sharply. This phenomenon is generally attributed to the belief that as wages rose, so did the service industries that provided the means for employees to spend their wages. Therefore, the service industries benefited from the economic growth and subsequently then played a larger part in the economic growth of the nation.

However, that would suggest that manufacturing and services are separate entities, when in reality we are seeing manufacturing create and consumer more and more services, year after year. These services, such as logistics, advertising, and marketing, make up an increasing amount of manufacturing time and expense. In the United States for example, for every $1 spent on manufacturing requires 16 cents on services. Furthermore, in some high-technology manufacturing, the number of employees involved in non-manufacturing roles, such as R&D, HR, office support, and IT, make up the large majority of employees.

Industry 4.0 places high importance on the Internet of services, where manufacturers can create or consume available services within their value chain. These services, such as inventory control, logistics, and smart transportation, will reduce costs, improve efficiency, and ultimately productivity. A critical challenge for manufacturers will be to approach the era of digital manufacturing in a more pragmatic way. Certainly manufacturers involved in heavily labor-intensive sectors such as jewelry, textiles and toys, will stay on the low-wage path. However, others in more automated lines must weigh-in factors such as access to low-cost transportation, to consumer insights, or to skilled employees. This change in strategy and production methods could result in a new kind of global manufacturing company. One that uses networked value chains that produce and consume services. Furthermore, it uses "big data" and analytics to respond quickly and decisively to changing global conditions in customer-insight and market conditions. Additionally, policymakers that recognize that their long-term goals for growth, innovation, and exports are through critical technical enablers such as investing in modern infrastructure will lead this new breed of manufacturer.

However, in order to adopt and take advantage of the enormous opportunities of the Industrial Internet and Industry 4.0, businesses will have to make some radical changes in every area of their business. The first and foremost prerequisite to adopting IIoT or Industry 4.0 is that of transforming the business to operate seamlessly in a digital world.

References

https://www.ge.com/digital/brilliant-manufacturing

http://www.airbusgroup.com/int/en/story-overview/factory-of-the-future.html

Digital Factories: The End of Defects - Industry & Automation - Pictures of the Future - Innovation - Home - Siemens Global Website

https://www.accenture.com/us-en/labs-insight-industrial-internet-of-things.aspx

https://www.accenture.com/us-en/insight-industrial-internet-of-things.aspx

https://theconsultantlounge.com/2015/07/accenture-report-internet-of-things-driving-new-era-of-living-services/

http://www.airbusgroup.com/int/en/story-overview/factory-of-the-future.html SAP - The Technical and Operational Solutions for Industry 4.0 in ERP

CHAPTER 15

Getting From Here to There: A Roadmap

In order for businesses to adapt to the concepts of the Industrial Internet, they need to realize the company's current position regarding their processes, procedures, philosophy, strategy, and current technologies in relation to the level of adaptation they wish to achieve. Then there is the thorny issue of the actual means to achieve that objective. A common question arises, "how do we get from where we are now to where we want to be"?

For companies to adopt and take advantage of the enormous opportunities of the Industrial Internet and Industry 4.0, they will have to make some radical changes in every area of their business. The first and foremost prerequisite to adopting an IIoT or Industry 4.0 strategy is that of transforming the business to operate seamlessly in a digital world. There are many approaches to digital transformation of the businesses, but the common consensus is that it involves addressing three key areas—customer experience, operational process, and business models.

Digital Transformation

The concept of *digital transformation*—which is the use of technology to improve performance—is as hyped as the Industrial Internet and is a hot topic of interest in the C-suite, and for good reason. Executives are seeing industries in all sectors benefiting from digitalization by using the latest digital technologies such as analytics, mobility, social media, and smart embedded devices.

Digital transformation is industry agnostic and is certainly not confined to technology startups, which may admittedly received much of the media limelight. Successful digital transformation projects are present across an array of modern and traditional industries, from banking to media to manufacturing. However just as digital transformation can affect all industries, it can also involve all core functions of a business, such as their strategy, processes, and procedures.

CapGemini, in conjunction with MIT Center for Digital Business, came up with the term digital transformation. They proposed an effective digital transformation program should look closely at not just what needed digitized, but how it could be accomplished.

Some of the key points from the CapGemini proposal was that business and IT must be closely aligned and work together to obtain digital transformation. They saw that the Business/IT relationship was key and both had to focus on the same challenges and ultimate goals for the program to succeed. This is extremely important because some proponents of digital transformation see it as a business led initiative that will be achieved with or without IT's blessing and participation.

However, that is a dangerous path to undertake, as IT in any business should always be aligned to the business strategy. Therefore, it is imperative that IT is fully engaged partners in the project or it will ultimately lead to disconnects and gaps in the program. Consequently, CapGemini recommended that the project be sponsored and championed from the top to ensure buy-in from all departments and stakeholders. Digital transformation is an extremely tricky project to manage, as it addresses core elements of the business, but neither customers, partners, or the competition are going to wait for change to happen. Change needs to be driven forward from the top.

CapGemini also suggested that opportunities in digital transformation existed in all industries. Furthermore, huge opportunities existed in efficiency, productivity, and employee enablement. However, to leverage those opportunities means transformation management intensity across all departments, stakeholders, and employees.

Furthermore, McKinsey suggests that digital transformation can reshape every aspect of the modern enterprise, with four core elements:

- Connectivity with customers and partners
- Innovation of products, business models, and processes
- Automation by replacing labor with technology
- Decision making by utilizing Big Data and advanced analytics

McKinsey proposed supporting these core activities and goals, with a cycle of functions of continuous improvement:

- *Customer experience*—Entails seamless multi-channel experience, and whenever, wherever service capability
- *Product and service innovation*—New digital products and services and co-creation of new products
- *Distribution, marketing, and sales*—Targeted higher returns on investment through optimized digital marketing and sales channels
- *Digital fulfillment*—Depends on automated processes and provisioning from start to finish
- *Risk optimization*—Focuses on embedded automation controls with risk profiling as well as by improving customer in-sight
- *Enhanced corporate control*—Improved real-time management systems and management dashboards, with seamless integration to the value and supply chains, including third parties

The big question is how a company shifts from where they are now to where they want to be through the digital transformation program. There are many roadmaps, which a company can follow on their journey toward digital transformation. However, not one roadmap fits all businesses or industries; each company's motivations, intent, goals, priorities, budgets, and pain points differ.

Additionally, most companies are not ready for full digital transformation and they only want to digitize certain areas of the business, such as addressing the customer experience or sales and marketing processes as they can reap early returns without too much risk of disruption to the core business. Another long-standing problem, especially in the traditional industries such as manufacturing, is the concept of functional silos and the need to work across those areas of functional and political segregation. For digital transformation to work, there must be minimum, preferably none, silos existing within the business. Unfortunately that is rather difficult in traditional industries that have a hierarchy of departments. For example, in manufacturing, the production departments—those that make money and are the sole reason for the company's existence—have seniority over the support departments such as IT or HR. Convincing these senior departments—such as manufacturing, engineering, logistics, finance, sales and marketing—to relinquish some of their status and influence is fraught with problems.

Ideally of course, all departments should cooperate for the greater good and learn from each other in order to boost productivity and profits. However, for that to work in practice, the CEO must understand the huge potential of digital transformation and how it will affect each area of the business as only he or she will be in a position to drive through the silos and create the ubiquitous business and technical connections required among people, processes, and systems.

Customer Experience

Customer experience, sometimes called user experience, although they are not always the same thing, is one of the basic building blocks of digital transformation and one that is typically addressed first. The reasons are simply because it can produce results quickly using existing equipment or by leveraging social media on cloud services.

Knowing the Customer

Social media has become a tremendous source of intelligence for businesses as it allows them to discover what leads to customer dissatisfaction. Additionally, companies are learning, some quicker than others, to embrace social media and engage with their customers/users directly. At the heart of these initiatives is the goal to know the customers better, to know their likes, preferences, and dislikes regarding the company's products or industry sector in general. If you can determine what customers' dislikes regarding the way the industry is run in general, there is a possibility to steal a march on the competitors. After all, just like all closed groups, companies in industrial sectors are often blind to their shortcomings. They think this is the way we have always done this, and so has everyone else, so they don't see a reason to change. However, getting constructive criticism from customers may be advantageous. An advantage of this is with the airline industry, before the advent of low-cost airlines, the customer experience was dreadful, checking-in was slow and tedious, as was boarding and the costs were high. However once a few pioneering airlines decided to cut costs they had to be sure that they would get enough passengers so they listened to their frequent flyers' pet gripes, and low and behold they revolutionized not just the pricing structures but check-in, boarding, and flight-booking procedures.

Customer Contact Points

One of the major digital contributions to the business is by the increase of customer contact points. Before mainstream digitization, back in the 2000s, companies could be contacted by telephone or e-mail. Today, companies have click to chat or call to communicate with a live support agent on their web sites, or they have online forms to open service tickets on their sites for customers' convenience. Other technologies are gaining acceptance, such as video calls using WebRTC, which enables a customer to contact a support center via a browser using video chat. This is proving important when technical support agents need to actually see the product and can make fault diagnosis and fault resolution much quicker.

Similarly, the massive growth in mobile apps has led companies to produce their own apps, which further integrates the customer with the company and provides even more intelligence such as location. These apps also can be used to send pictures or videos of a product to the customer. Similarly a customer can send media to the support center in order to assist a technician with a diagnosis.

Transforming Operational Processes

Customer experience may be the first channel that a company turns to digitize because of its importance and relative ease of implementing the new technologies. However, when executives are asked about the most successful and productive areas to be digitized, they tend to mention process digitization. The digital transformation of operational processes can reap very early benefits, sometimes termed picking the low lying fruit, so it is an attractive initial area to focus on.

Process Digitization

The benefits of process digitization extend right through the value chain from early rapid development and prototyping of a product, to automated production lines and efficient stock control and dispatch. Automation of production line enables workers to be freed to do other less tedious repetitive jobs, such as supervising automated processes and using their production and product experience in a quality-control capacity.

Digitizing processes also saves money as products and stock are more efficiently created and replenished using automatic stock replenishment procedures in ERP. Digitization facilitates a variety of stock handling and inventory controls, such as build-to-stock, build-to-order, or engineer-to-order. By moving away from the build-to-stock model to a much more cost-effective and efficient build-to-order can have a huge effect on the asset cost of inventory and on raw materials and parts.

Worker Mobility

Mobility is one of the great enablers of efficiency and productivity in the last decade. With the ubiquitous smartphone, employees are now mobile and initiatives such as BYOD (bring your own device) and even BYOC (bring your own cloud) have allowed employees to be mobile and work from anywhere at any time. This has greatly improved productivity and innovation. Similarly, VoIP PBXs have allowed employees to work from anywhere as calls from the company phone system can be redirected to any phone seamlessly and transparently to the caller. Therefore, employees can take calls when travelling, at home, or in the car, just as if they were at their desks.

VoIP PBX systems also seamlessly integrate with the Internet so employees can participate in conference video calls and remote meetings and presentations. Additionally, they can collaborate within these calls by sharing their desktops and using them as shared whiteboards, or by running a presentation. The opportunities for remote presentations and teaching are vast, saving companies not just money on travel and wasted time, but boosting collaboration and innovation across disperse groups of experts.

Performance Management

Performance and operational efficiency gains are the results of digital transformation. Digital information collated and presented as performance and KPI indicators in a dashboard format enables executives and managers to see exactly what is happening within the business in real time. This would have been impossible for most organizations a decade ago where some still relied on departmental status reports produced on Excel spreadsheets or unconnected propriety software.

Advances in system integration techniques, such as APIs (application programmable Interfaces) and web services, and particularly the advent of SaaS (software as a service), has greatly enhanced the ability to retrieve report data from disparate systems. This allowed businesses to collate and present operational and management data either as a dashboard or as the input to other processes.

Big Data and advanced analytics take this concept even further, as previously the analytics was historically based, with some vague trend analysis. However, in a digitized environment, data analysis software collects and analyzes data in real time. This facilitates the production of plan-to-performance analysis, which allows management to have visualization of all assets, projects, business units, and employees that they manage to see if they are performing as expected against their business goals. Plan-to-performance analysis will highlight not just the overall performance status but also provide granular drill-down reports on sub-projects or procedures to aide understanding of why the status is as it is. Furthermore, with Big Data's advanced analytic capability, data analysis can now be historical, predictive, and prescriptive.

Transforming Business Models

Many business executives have enough insight into their own business and that of their industrial sector that they operate successfully. It is actually common to have managers who have enough self-awareness that they understand that change is imperative if the business is to survive, let alone grow. However, with digital transformation and the Industrial Internet, companies can change and adapt to new innovative business models.

Digitally Modified Business

The problem is that modifying a business strategy can lead to dire consequences. Take as an example a solid business that has been delivered through traditional means. An example could be a butcher shop that has delivered meat and cut joints as requested every week. However, the butcher learns about digitization and e-commerce and, staggered by the success stories, he decides to stop selling locally and sets up an online shopping market instead.

Strangely, we did see these ludicrous success stories in the media back in the late 1900s as local butchers claimed to produce and sell vast amounts of local pies, pasties, lamb joints, and other foodstuff on the Internet, regardless of logistics and production restraints.

However, the point is that modifying a business model is not always a good idea, and in fact it can be a very bad suggestion. The butcher's current business model appeared to be successful, traditional, and well accepted, so why then would he want to change it?

To understand this, let's look at how problems can occur.

New Digital Business

In order to create a new business or develop a new idea, you need innovation and tremendous amount of imagination, perseverance, and dedication. In the case of the butcher starting an online business, he would need to ensure that his current customer base would follow him and be positive about the change in direction. After all, the butcher's customers may trade with him because of his traditional methods, as perhaps they also are not digitized, or even have that future capability. It would be inevitable that the butcher would lose more customers than he would gain by a complete digital transformation. The butcher would be far better to run the traditional business in parallel with a new digital business initiative, at least until he could gain experience in the nuances of online marketing by spending time and energy building an identity, a new customer base, and product lines suitable for trading in an online world.

Chapter 15 | Getting From Here to There: A Roadmap

The problem of course is even more profound when you're launching new products or services and especially when launching a new business or entering a new digital market. The problem is that it is not easy to predict how customers will react and a startup company will inevitably struggle to deliver the goals and the goods at the outset until they build an identity and reputation. Therefore, it is not always clever to change the business model or technology just for the sake of it. Let's look at one real-world scenario as an example.

A technological company that was unicorn rated at $2.8 billion as a startup, failed to live up to expectations and after a few years in development, having never traded, crashed and went bankrupt. This turn of events came about simply because the company failed to deliver a consistent product. The company was involved in mobile payments and it had a great concept, reasonable hardware, and a large niche market to fill—in mobile phones accepting credit card payments. However, the company that once claimed it would be larger than Google and Alibaba went bankrupt simply because it had no stable product as it kept switching to the latest technology—it was never happy with the technology that it was using. The problem, however, was not upgrading the technology per se, it was that each time the company had to embark on another mass marketing and advertising campaign to push this new technology, losing more potential customers than they gained along the way.

However, saying that, it is always sensible to look at alternatives and other business options and that is what makes digital business so popular. By analyzing data and performance figures, a company can ascertain true potential figures and trends, and thereby derive reliable trend analysis.

Digital Globalization

The whole point of digitization is that industrial industry and Industry 4.0 can span global networks. The global effect, that sense of collaboration and team building across borders, makes the IoT viable. Consider for a moment how industry could be feasible if each division of industry did its own thing.

If we consider that industry depends on true analytics, procedures, and manufacturing processes, we can say that these goals will produce an ultimate product. However, it is not always that way. If we analyze the data, we can see that the process of data to information, and then to knowledge, is not sometimes clear.

This is why we must consider collating data across a vast global environment in order to aggregate and then use that data lake as a pool for analysis. The larger the data lake, the more likely our analysis is going to be.

In order to meet global data acceptance, we must accept that the global data pools are not only trustworthy but essential in order to derive information and knowledge.

Increase Operational Efficiency

Consultants often stress the point that the whole purpose of the Industrial Internet or Industry 4.0 is about increasing industrial operational efficiency. However, that is only partially true. Industrial and business projects are targeted at providing efficiency, automation, and profits across the entire supply chain.

However, they are correct to stress the importance of operational efficiency as it is paramount to all business and Industry 4.0 lends itself to increased productivity, efficiency, and customer engagement.

Merge OT with IT

The biggest problem with merging OT (operational technology) with IT (Information technology) is that they have completely different goals and aspirations. It is actually similar to merging operations and development into devops. In reality, OT is about manufacturing and OT workers and technicians have evolved via a different mindset. OT workers have come through the industrial workforce, where employees are labor-oriented and expect that the job they do is vital to the manufacturing of the product.

OT staff work hard in difficult conditions and they work to meet production targets and work closely with the factory workforce as part of a team.

IT, on the other hand, is much more suited to the enterprise and the business and they use their expertise to guide other departments, to use efficient and productive methods and technologies. IT tends to lead rather than collaborate and that can cause stress, but either way the integration of OT and IT is hugely important to the business. Therefore, it is vital to first plan the convergence of OT and IT with an initial pre-convergence stage. During this pre-convergence stage, it is important to use internationally acceptable standards and to identify the company strategy so that there is alignment planned.

Once IT and the business have agreed on a convergence strategy, the actual process of converging OT and IT can begin. Convergence will be considered to be complete upon certain stage goal and milestones being achieved. These typically point toward a converged infrastructure where every device has an IP address and fall under a centralized network-management system. Once all the devices in the network are under the joint management of OT and IT, there is scope for collaboration in development and maintenance.

The next step after convergence is alignment. We touched on this earlier. IT alignment with the business is best practices and IT must ensure that it align its strategy and tactics to the company's business strategy. We can consider the company aligned at an engineering and technical level when devices and systems are remotely accessible via the Internet. Furthermore, it is ideal if engineering and IT collaborate to chart all the applications and informational sources, such as disparate databases. Furthermore, it is best practice for IT to integrate all the applications, systems, and data sources with a common enterprise-wide identity and access management system.

The final step is to build on the systems' alignment by integrating them all under the one planned architecture. It is at this stage that the company realizes cost savings, operational efficiencies, and competitive advantage.

The main point is that when merging the departments of OT and IT, management must carefully plan each stage before taking action; this is something that OT would find more natural than the more opportunistic IT.

Increase Productivity via Automation

An essential part of automation in manufacturing is to remove where possible any human action or interaction from the production process. A prerequisite to achieving this goal is that process controllers to systems, machines, and appliances can assign processing tasks. This is termed M2M, machine-to-machine communication, and within the context of human-machine-interaction, it is a vital component of the smart factory as it forms the cyber-physical systems. The CPS communicate through the Internet and, via the Internet of Things and services, produce new plant models and improves overall equipment effectiveness (OEE).

However, it is not just in industrial processes where M2M are commonplace, as they are ubiquitous throughout many business processes and indeed in any process where networked smart devices have a role in the process chain.

The networking of these digital things will also provide a huge spinoff for telecom companies and Internet service providers who will have to provide the traffic transportation between devices. Indeed, telecom companies are predicting huge increases in the number of SIMS and data modems integrated into all sorts of remote devices, such as vending machines, connected cars, trucks for fleet management, smart meters, and even remote health monitoring equipment, by 2020.

Automation is the way forward and, as we have just seen, it relies heavily on effective M2M in the process chain. M2M should play a large part in the business convergence and digital transformation process, as it not only improves productivity through overall equipment effectiveness but also allows for new and innovative business models.

Develop New Business Models

Industrial companies create their business models based on competitive strategy, which involves business differentiation, cost leadership, and focus. In most industries, especially in manufacturing, this strategy still holds true. However, with the advent of digitization and connectivity came new ways of looking at traditionally sound strategies in creating and capturing value.

As management shifts their focus toward digitization and perhaps a further evolution toward Industry 4.0, they should become aware of the huge opportunities for innovation to regard to value creation and value capture. Cloud-based services and techniques have enhanced the potential of value creation and capture to such a level that existing business models will require a rethink.

At the heart of any company's business model or strategy is value creation, as it is the sole reason that most businesses are in existence. Value creation is about increasing the value of a company's offerings—products or services—that encourage customers to pay for them or utilize them in some way beneficial to the business. In manufacturing and the product-focused business, creating value historically meant producing better products with more features than the competition. This required that businesses identified enduring customer needs and that they fulfilled that through well-engineered solutions. Of course, other businesses would be striving to fulfill the same customer needs so competition would ensue based on features, quality, and price. The strategic goal was to create and sell products with the hope that once the product became obsolete the customer would buy a replacement.

However, the Industrial Internet has presented an opportunity to revolutionize the way that businesses can create and sell products. There is no longer any excuse for the one-and-done product lifecycle, as manufacturers can track customer behavior and offer over-the-air updates, new features, and functionality throughout the product's lifecycle. Furthermore, products are no longer in isolation. With the advent of the Internet of Things, connectivity is king and products can interact with other products. Connectivity leads to new insights and products through analytics, which improves forecasting, process optimization, product lifecycle support, and a better customer experience.

Consequently, modern business models are focusing on the customer, by creating value of experience. The Internet of Things facilitates business to view the customer's experience in new ways, from how they initially view the product, how they use it, and what it connects with and ultimately to learn what more the product could do or what services or features could revitalize the product. Additionally, making money from the product is no longer restricted to the initial sale, as now there is potential for other revenue streams such as value-added services, subscriptions, and apps.

Similar to value creation, the way that businesses capture value has changed with the advent of cloud services, which leads to the monetization of customer value. Traditionally, at most product-driven businesses, value capture has simply been about setting the right price to maximize profits on discrete product sales. Of course, that is a simplistic view, as most companies expend a great deal of energy and creativity presenting and marketing their products and searching for key differentiators from the competition.

However, businesses can now maximize margins and leverage their core competencies to bring a product to market. Furthermore, they can do this while controlling the key points in the value chain, such as commodity costs, brand strength, or patents. They can also add personalization and context to lock in customers, which leads to recurring revenue.

Adopt Smart Architectures and Technologies

Innovation is critical in developing new business models and opportunities. However for companies to be able to fully exploit the opportunities they will have to master three core competencies—sensor-driven computing, industrial analytics, and intelligent machine applications.

Sensor-Driven Computing

Sensor-driven computing is the basis of the Industrial Internet as sensors provide the connection between the analogue world of our environment such as temperature, voltage, humidity, and pressure and the digital world of computers. Sensors provide objects with perception into their state and their surroundings and they provide the data required by systems to gain insights into industrial processes. Sensors only supply raw data to gain actionable insights and analytics.

Industrial Analytics

Industrial analytics converts raw environmental data collected from perhaps thousands of sensors into human understandable insights and knowledge. Analytics, traditionally, due to the limits of technology, had a focus on historical data, such as monthly sales reports. However, with the advent of cloud computing and mass data storage, advanced analytics has become commercially available to everyone. Advanced analytics now provide industry with historical, diagnostic, predictive, and even proscriptive analytical data. These advanced analytical algorithms provide insights into not just what has happened, but why it happened, when it might happen again, and what you can do about it.

Intelligent Machine Applications

Analytics have profound importance in industrial scenarios, as they provide the actionable insights that facilitate intelligent process control and proactive decision-making. However, to leverage the proactive benefits of predictive analytics requires intelligent machines, ones that are not just mechanical but have built-in intelligence. These smart machines will have self-awareness, not in philosophical terms, but an awareness of their own and their process' current state through self-diagnostics. Being able to predict events in regard to component failure provides the methods to move from break-fix to fix-before-failure, which has profound economical benefits to industry. However, the real benefit of having intelligent machines is that they can integrate and collaborate with one another across domains. This enables developers to use innovation when creating intelligent applications.

Reaping the optimal benefits of intelligent connected technology requires a strategic rethink, technical awareness, and innovation. However, all that creativity must be based on a robust technical architecture and infrastructure, which requires an IIoT platform. The Industrial Internet platform is still at a level of immaturity so there are still gaps in interoperability and information sharing. Currently, this is the overriding technical challenge to businesses wanting a roadmap to the Industrial Internet.

Transform the Workforce

Back in the 70s, there was major concern among business leaders that production line automation with robots would replace the work performed by humans and effectively render them redundant. The problem was and still is that employees are the heart and soul of a company, unless of course you are operating a lights-out manufacturing facility. At the time, CEOs claimed that reducing the labor-intensive workforce from tedious, dirty, boring, or dangerous work was beneficial to the employee to the business. These business leaders managed to convince themselves that an automation initiative was humane and economical. Furthermore, it was an efficient way to boost productivity and efficiency while reducing costs and boosting bonuses. Unsurprisingly, trade unions and those whose jobs and livelihoods were at risk strenuously objected to this strategy, pointing out it wasn't just them that were at risk.

Although it might have been attractive for CEOs at the time to reduce the payload and the operational expense and offload low-skilled workers, while investing in skilled IT generalists who could perform a variety of task, the premise was flawed.

Paul Krugman, back in 1996, imagined a scenario where:

> "Information technology would end up reducing, not increasing, the demand for highly educated workers, because a lot of what highly educated workers do could actually be replaced by sophisticated information processing —indeed, replaced more easily than a lot of manual labor."

Paul Krugman's words have proven to be profound as we are now seeing automation replacing not just casual labor but highly skilled workers where market forces have seen skilled jobs replaced by software. Careers in software development and programming were once, even in 2012, promoted by universities and colleges as the work of the future, when they are now in the front of the automation queue.

It is an immutable truth that the labor workforce will reduce but the business will also have to transform in order to meet the requirements of the digital connected age. Businesses will require business analysts, strategists, data scientists, and those skilled in developing algorithms that match company strategy. It is one thing to collect vast amounts of operational data, but if you cannot articulate the correct questions and make sense of the returned answers, it is worthless. Consequently, reducing the head count of low-paid manual workers will be operationally beneficial in the short term, but any short-term benefit will be overwhelmed by the costs of expert hires as the company transforms to the digital age.

Index

A

Access network
  carrier Ethernet, 122
  Ethernet, 120
  I/O data, 124
  IP routing, 122
  MPLS, 122
  operational and
    management domain, 120
  Profinet, 123–124
  VLAN, 120
Advanced message queuing
  protocol (AMQP), 137
Analogue-to-digital convertor (ADC), 187
Application programming interface (API)
  analogy, 145
  application programmer, 145
  businesses create apps, 146
  component parts, 144
  database designer, 145
  DBA, 146
  external application, 147
  microservices, 151
  open APIs, 147
  preformatted template, 146
  REST
    caching, 150
    HTTP verb binding, 151
    programs, 149
    security, 151
    standardization, 148
    URL, 149
  vs. SOAP, 150
  web and mobile applications, 147
  security, 151
  SOA, 21, 143–144
  SOAP
    built-in error handling, 149
    definition, 147
    HTTP verb binding, 151
    standardization, 148
    support modules and options, 148
    vs. REST, 150
    WSDL, 148
    XML, 148–149
  SQL query, 146
  URLs, 146
Assistive technology, 15
Augmented reality (AR), 59

B

Baymax, 15
Building management
  systems (BMS), 21
Business-to-consumer (B2C), 2

C

Carrier Ethernet, 122
CMS, 22
Commodity off-the-shelf (CotS), 42
Common object request broker
  architecture (CORBA), 148

Constrained application
protocol (CoAP), 128
Control area network (CAN), 181
Customers' premise equipment (CPE), 42
Cyber-physical system (CPS), 36

D

Data bus, 139
Data distribution service (DDS), 138
Data management, 82
Delay tolerant networks (DTN), 139
Distributed component object
model (DCOM), 148
Dynamic name server (DNS), 127

E

Epidemic technique, 141
Ethernet, 120, 127
Extensible Messaging and Presence
Protocol (XMPP), 137

F

Functional domains, 69
  asset management, 71
  communication function, 70
  control domain, 70
  executor, 71
  modeling data, 71

G

Giraff, 15
Google Glass, 25

H

Human machine interface (HMI), 20–21, 45
HVAC system, 131

I, J, K

Identity access management (IAM), 191
IIoT architecture
  architectural topology, 75
  data management, 82

advanced analytics, 84
queries, 83
storage, persistence,
  and retrieval serves, 83
IIAF
  application domain, 75
  Business domain, 75
  Business viewpoint, 68
  functional domains
    (see Functional domains)
  information domain, 73
  operation domain, 72
  stakeholder, 67
  usage viewpoint, 68
implementation viewpoint, 75
Industrial Internet
  IIC, 66
  IISs, 66
  ISs, 66
  M2M, 66
key system characteristics, 79
  communication layer functions, 81
  connectivity functions, 80
  data communications, 79
  deliver data, 80
M2M, 65
three-tier topology
  communication transport layer, 78
  connectivity, 78
  connectivity framework layer, 78
  edge tier, 76
  enterprise tier, 76
  gateway-mediated edge, 77
  platform tier, 76
IIoT middleware
  architecture, 156
  commercial platforms, 160
  components, 156
  conceptual diagram, 154
  connectivity platforms, 157
  mobile operators, 158
  open source solutions, 160
  requirements, 159
IIoT WAN technology
  3G/4G/LTE, 164
  cable modem, 166
  DWDM, 165
  free space optics, 166

Index

FTTX, 165
internet connectivity, 162
M2M
    Dash7 protocol, 172
    LoRaWAN architecture, 171
    LTE cellular technology, 175
    MAC/PHY layer, 169
    millimeter radio, 176
    OSI layers, 169
    requirements, 167
    RPMA LP-WAN, 173
    SigFox, 170
    Weightless SIG, 175
    Wi-Fi, 174
MPLS, 164
SDH/Sonnet, 163
VSAT, 167
WAN channels, 162
WiMax, 166
xDSL, 163
Industrial Internet
    3D printing, 60
    augmented reality (AR), 59
    Big Data, 52
        business value, 55
        variety, 54
        velocity, 54
        veracity, 55
        visualizing data, 55
        volumes of, data, 53
    CAN network, 181
    Cloud model, 47
    CPS, 35
    fog network, 51
    ICS, 180
    IFE, 182
    IP Mobility, 40
    M2M learning and artificial intelligence, 56
    Miniaturization, 34
    Network virtualization, 43
    NFV, 42
    people vs. automation, 62
    remote I/O devices, 34
    Russian hackers, 180
    SDN, 44
    SDN vs. NFV, 45
    security
        CAN bus, 192
        Ethernet/IP, 189
        IAM, 191
        ICS-CERT, 191
        IOT network, 183
        IP and Ethernet, 184
        Modbus, 189
        OT network, 183, 187
        OT vs. ICS, 185
        PCL and DCS, 187–188
        physical and behavioral security, 186
        ping devices, 185
        PLC, 183
        Profibus, and Profinet, 189
        system level, 190
        VHF radio equipment, 192
        VLAN network, 189
        Y2K bug, 185
    smartphones, 45
    Ukraine power, 180
    Wireless communication technology, 38
Industrial Internet. See Industrial internet of things (IIoT)
Industrial operations technology (IOT), 1–2, 183
Industrial internet architecture framework (IIAF), 67
Industrial internet consortium (IIC), 66
Industrial internet of things (IIoT)
    B2C, 2
    Big Data, 3, 5
    building's energy efficiency, 20
    business gains, 3
    catalysts and precursors
        adequately skilled and trained staff, 6
        innovation, commitment to, 6
        security, 7
    cloud-computing model, 6
    commercial market, 1
    consumer market, 1
    digital and human workforce, 11
    digital twin, 11
    green house gas emissions, 19
    heath care, 14
    Industry 4.0, 2
    innovation, 7
    installing sensors and actuators, 20
    intelligent devices, 8
    IOT, 1–2
    IOT, disadvantages, 20

Index

Industrial internet of things (IIoT) (cont.)
    IOT6 Smart Office, 21
    IT sectors, 5
    key opportunities and benefits, 8
    logistics
        adopting sensor technologies, 24
        advanced telemetric sensors, 26
        augmented reality glasses, 25
        automating stock control task, 24
        barcode technology, 23
        Big Data, 26–27
        document scanning and verification, 26
        forklift, 24–25
        Google Glass, 25
        multiple sensors, 26
        pick-by-paper, 25
        RFID, 23–24
        SmartLIFT technology, 24–25
        temperature and humidity sensors, 24
        track and trace, 26
    M2M, 3
    manufacturers, 10
    Oil and Gas industry
        automated remote control topology, 18
        automation, 18
        Big Data analytics, 19
        cloud computing, 17
        data analytics, 16
        data collection and analysis, 18
        data distribution system, 17
        DDS bus, 18
        down-hole sensors, 16
        drilling and exploration, 16
        industry regulations, 16
        intelligent real-time reservoir management, 19
        interconnectivity, 17
        MQPP and XMPP, 17
        remote node's status, 17
        6LoWLAN and CoAP, 17
        technological advances, 16
        wireless technologies and protocols, 17
    outcome economy, 10
    power of 1%, 4
    retailer
        innovations, 29
        IT costs, 27
        POS, 27–28
        real-time reporting and visibility, 28
        stock control, 28
    sensor technology, 4
    smartphone, 20
    WSN, 21
    WWAN, 5
Industrial Internet system
    communication protocols
        Ethernet protocol, 100
        industrial Ethernet, 98
        TCP/UDP containers, 100
    concept of, IIoT, 88
    diverse technology, 116
    gateways, 115
    heterogeneous networks, 116
    industrial gateway, 118
    industrial protocols
        current loop, 97
        field bus technology, 98
        RS232 serial communications, 96
    proximity and access network
        address types, 114
        IIoT context, 115
        IPv4, 109
        IPv6, 112
        IPv6 Subnets, 114
        NAT, 111
    proximity network, 89
    wireless communication technology, 102
        bluetooth low energy, 103
        IEEE 802.15.4, 102
        NFC, 107
        RFID, 106
        RPL, 108
        6LoWPAN, 107
        Thread, 107
        Wi-Fi backscatter, 105
        ZigBee, 103
        ZigBee IP, 104
        Z-Wave, 105
    WSN edge node, 90
        functional layers, 93
        IP layers *vs.* IIoT layers, 95

Index

low-power technology, 91
network protocols, 91
OSI table, 93
web 2.0 layers, 94
Industrial Internet systems (IISs), 66
Industrial systems (ISs), 66
Industry 4.0
  advantages, 199
  big data and analytics, 208
    additive manufacturing, 210
    architecture, 211
    augmented-reality-based systems, 210
    business processes, 213
    cloud data, 210
    customer acceptance, 215
    customer evaluation, 214
    cyber-security, 210
    equipment, 212
    horizontal and vertical system integration, 209
    IOT, 209
    products, 213
    simulation, 209
    smart manufacturing, 211
    supply chains, 213
    use of, robots, 209
    workforce, 212
  characteristics, 199
  cyber-physical systems, 196
  definitions, 197
  design principles
    decentralization, 207
    interoperability, 207
    modularity, 208
    real time capability, 208
    services, 208
    virtualization, 207
  dynamic process control, 196
  global networks, 195
  manufacturing processes, 196
  value chain, 201
    business benefits, 205
    Cost-cutting, 203
    creation, 203
    horizontal activities, 201
    quality, features, 203
    support function, 202

In-flight entertainment (IFE), 182
Internet of Things (IOT), 1–2, 29
IOT6 Smart Office, 21
IP, 126
IPv6, 21, 23

L

Late-binding, 133

M

M2M learning and artificial intelligence, 56
Machine-to-machine (M2M), 3, 6
Message bus, 132
Message queue telemetry transport (MQTT), 136
Micro-electro-mechanical systems (MEMs), 53
Microservices, 151
Mobile device management (MDM), 158
Multiprotocol label switching (MPLS), 122

N, O

Near field communication (NFC), 20, 107
Network address translation (NAT), 111
Network functionality virtualization (NFV), 42

P

Point of sales (POS) machines, 27–28
Profinet, 123–124
Programmable logic controls (PLCs), 183, 224
Proof-of-concept (PoC), 35
Prophet, 141
Publish/subscribe protocol, 133

Q

Quality of service (QoS), 122, 138

Index

R

Radio frequency identification (RFID), 20
Real-time reaction, 132
Reliable transport protocol (RTP), 128
Remote operational vehicles (ROV), 9
Return on investment (ROI), 68
RFID, 22–24, 29
Road map
    business models
        digital globalization, 238
        digitally modified business, 237
        new business, 237
    customer experience
        contact points, 235
        customers/users directly, 234
    digital transformation, 232
        CapGemini proposal, 232
        core elements, 233
        goals, 233
    operational efficiency
        create new business, 241
        merge OT, IT, 239
        via automation, 240
    operational processes
        mobility, 236
        performance management, 236
        process digitization, 235
    smart architectures
        industrial analytics, 242
        intelligent machines, 243
        sensor-driven computing, 242
Rolls Royce jet engines, 11

S

Sensor's management device, 132
Sensor technology, 4
Service-oriented applications (SOA), 21, 143–144
Simple open architecture protocol (SOAP)
    built-in error handling, 149
    definition, 147
    HTTP verb binding, 151
    security, 151
    standardization, 148
    support modules and options, 148
    vs. REST, 150
    WSDL, 148
    XML, 148–149
Smart factory
    Airbus, 224
    asset efficiency, 222
    automated processes, 226
    computers handling, 225
    defects, 225
    employees payments, 227
    EWA, 224
    GE Brilliant Factory, 223
    Industry 4.0, 225
    manufacturing and services, 229
    manufacturing processes, 218
    mobile app, 223
    production line diagram, CPS, 219–220
    RFID tags, 218
    SERP system, 221
    Siemens Chengdu, 227
    winners and losers, 222
SmartLIFT technology, 24–25
Systems-on-a-chip (SoC), 35

T

TensorFlow, 63
3D printing, 60
Transport control protocol (TCP), 126

U

Unreliable data protocol (UDP), 127, 129, 134, 139

V

Very small aperture terminal (VSAT), 167
Virtual LAN (VLAN), 120
VLAN Trunking Protocol (VTP), 121

W, X, Y, Z

Web services description language (WSDL), 148
Wireless sensor networks (WSN), 21
Wireless wide area networks (WWAN), 5

Get the eBook for only $5!

Why limit yourself?

Now you can take the weightless companion with you wherever you go and access your content on your PC, phone, tablet, or reader.

Since you've purchased this print book, we're happy to offer you the eBook in all 3 formats for just $5.

Convenient and fully searchable, the PDF version enables you to easily find and copy code—or perform examples by quickly toggling between instructions and applications. The MOBI format is ideal for your Kindle, while the ePUB can be utilized on a variety of mobile devices.

To learn more, go to www.apress.com/companion or contact support@apress.com.

All Apress eBooks are subject to copyright. All rights are reserved by the Publisher, whether the whole or part of the material is concerned, specifically the rights of translation, reprinting, reuse of illustrations, recitation, broadcasting, reproduction on microfilms or in any other physical way, and transmission or information storage and retrieval, electronic adaptation, computer software, or by similar or dissimilar methodology now known or hereafter developed. Exempted from this legal reservation are brief excerpts in connection with reviews or scholarly analysis or material supplied specifically for the purpose of being entered and executed on a computer system, for exclusive use by the purchaser of the work. Duplication of this publication or parts thereof is permitted only under the provisions of the Copyright Law of the Publisher's location, in its current version, and permission for use must always be obtained from Springer. Permissions for use may be obtained through RightsLink at the Copyright Clearance Center. Violations are liable to prosecution under the respective Copyright Law.

Printed in Great
Britain
by Amazon